WHAT IN THE WORLD IS DIABETES?

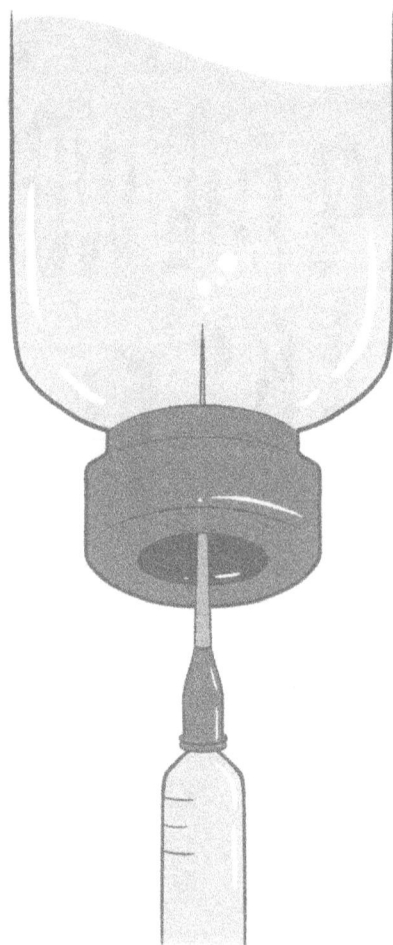

WHAT IN THE WORLD IS DIABETES?

Written by:
Austin Mardon
Aleefa Devji
Janani Rajendra
Susie Woo
Tara Y.T. Chen
Alicia Au
Sudipta Samadder
Ayah Nour Nehdi
Maria Gonzalez
Eline El-Awad Gonzalez
Megan Ng

Edited by:
Naima Mohamood

GM PRESS

2021

A Golden Meteorite Press Book.
Printed in Canada.

First Printing: 2021

Typeset and Cover Design by Fariha Khan

Email: aamardon@yahoo.ca
Telephone: 1-(587)-783-0059
Website: www.goldenmeteoritepress.com

Additional copies can be ordered from:
Suite 103 11919 82 Street NW
Edmonton, AB
T5B 2W4
CANADA

ISBN: 978-1-77369-259-3

TABLE OF CONTENTS

1

THE HISTORY OF DIABETES

By Alicia Au

INTRODUCTION

Diabetes mellitus is a disorder that has been known since antiquity. Descriptions of it have been found in the Egyptian papyri, in the literature of ancient Greek and Arabic physicians in addition to Indian and Chinese medical literature (Karamanou et al., 2016). The timeline of important discoveries of diabetes goes as follows:

- In the second century AD, Aretaeus of Cappadocia is credited with the first accurate description of diabetes, creating the term "diabetes" (Laios et al., 2012).
- In the 17th century, Thomas Willis added the term "mellitus" to the disease to describe the sweet taste of the urine and to distinguish it from diabetes insipidus, meaning tasteless (Barnett, 2010).
- In the 19th century, Claude Bernard discovered the glycogenic function of the liver which showed the causation of diabetes mellitus (Ahmed, 2002).
- In 1889, Oskar Minkowski and Joseph von Mering performed an experiment involving dogs illustrating the pancreatic function of producing insulin, a hormone which controls blood sugar levels (Luft, 1989).

1

- In 1921, Frederick Banting and Charles Best successfully built on Minkowski's and Mering's experiment by isolating insulin from pancreatic islets and administering it to patients suffering from type 1 diabetes (Simoni et al., 2002).

Although diabetes mellitus has been known since antiquity and despite therapeutic advances, it is still an incurable chronic disease today. In this chapter, prominent steps in the history of diabetes mellitus from antiquity until present will be highlighted to illustrate the development of the current knowledge in diabetes mellitus.

WHAT IS DIABETES AND WHY IS IT IMPORTANT?

Diabetes is a group of chronic metabolic diseases impacting the body's ability to use food properly, specifically impacting carbohydrate, lipid and protein metabolism (Zimmet et al., 2014). It is distinguished by long-term high blood sugar levels as a result of a problem in insulin secretion, insulin function or a combination of both (Karamanou et al., 2016). There are 2 main types of diabetes mellitus: type 1 diabetes which is insulin dependent, and type 2 diabetes which is non-insulin dependent (Karamanou et al., 2016). Type 1 diabetes is caused by the autoimmune destruction of the beta cells of the pancreatic islets whereas type 2 diabetes is caused from insufficient insulin secretion and resistance to the function of the insulin (Karamanou et al., 2016). Statistics show that 9% of adults have diabetes mellitus while it was estimated that in 2012, 1.5 million people died due to the disease (Karamanou et al., 2016). According to the World Health Organization (WHO), diabetes will be the 7th leading cause of death in 2030 (Karamanou et al., 2016).

ARETAEUS OF CAPPADOCIA

Aretaus is one of the greatest physicians of the Greco-Roman antiquity (Laios et al., 2012). He was born in Cappadocia, a region in modern day Turkey, studied medicine in Alexandria, Egypt and practiced in Rome in the second century AD (Laios et al., 2012). At this time, medical practice was based on the principles of the Pneumatic school believing in the vital role of pneuma (air) but also embracing the theory of the four humours (hear, colness, moisture and dryness) (Karamanou et al., 2016).

Before aretaeus, ancient Greek medical authors mentioned that diabetes caused excessive thirst, polyuria, emaciation of the human body, leading sometimes to death. The symptom of polyuria gave the idea to Galen, one of the Greek medical authors, to name the disorder diarrhea urinoma (diarrhea of the urine) (Karamanou et al., 2016). Later, the term diabetes was introduced into medical vocabulary by Aretaeus. It comes from the Greek verb διαβαινω (diabaino) which means "I pass through" and which describes diabetes, the condition in which the fluid runs through (Karamanou et al., 2016).

THOMAS WILLIS

The English anatomist and physician Thomas Willis, is considered one of the greatest physicians in the 17th century (Karamanou et al., 2016). He studied classics and then medicine at Oxford where he was appointed Professor of Natural Philosophy (Karamanou et al., 2016). During his career, he greatly contributed to the literature of the anatomy of the brain and nervous system, based on his own dissections, and remains very celebrated (Karamanou et al., 2016).

He commented on the sweet taste of the urine in diabetic patients, adding the term mellitu (Karamanou et al., 2016). It was actually a rediscovery, as in the 7th century BC, the Indian physician Shuruta mentioned the sweetness of the urine but this work was unknown to Willis so, he was the first European medical writer who mentioned it (Karamanou et al., 2016).

CLAUDE BERNARD

Born to a poor family in the south of France, because Bernard at the age of 19 was an apprentice to an apothecary (Karamanou et al., 2016). His passion for the theater led him to write two plays but he was discouraged by the literary critic and politician Saint-Marc Girardin who encouraged him to enroll in medicine (Karamanou et al., 2016). Bernard's research career was a success. In 1854, he became a member of the Academy of Sciences and later on he succeeded to the chair of experimental physiology at the College de France (Karamanou et al., 2016). It was recounted that Emperor Napoleon III admired him so much that created two laboratories for him (Karamanou et al., 2016).

Bernard's contribution in the study of metabolism and diabetes is still relevant today. In the 19th century, scientists were trying to figure out the role of pancreas in the physiopathology of diabetes (Karamanou et al., 2016). They found clues in the post-mortem examinations of the diseased, atrophic or stone filled pancreases (Karamanou et al., 2016).As they believed that the pancreas was an exocrine organ, they were mislead to interpret these findings as a phenomenon by chance (Karamanou et al., 2016). However, Claude Bernard decided to test this hypothesis.

At the beginning, he was misled to believe that "diabetes was a nervous affection of the lungs" (Karamanou et al., 2016). However, during an experiment, he injected grape sugar into the vein of a dog, extracting at the same time blood from an artery (Karamanou et al., 2016). This blood contained high concentration of sugar and he realized that glucose was not destroyed in the lungs, because blood must pass by these organs in order to move from the jugular vein to the carotid artery (Karamanou et al., 2016). He was then fed dogs on a carbohydrate-rich diet, the blood from the hepatic veins and vena cava contained sugar which was not destroyed in the liver and was also present in heart ventricles, so the theory of the lungs' role in diabetes was rejected (Karamanou et al., 2016). In further experiments, Bernard proved that animal blood contains sugar even if it is not supplied by food (Karamanou et al., 2016). Testing the theory that sugar absorbed from food was destroyed when it was passing through tissues, Bernard put dogs on a carbohydrate diet and killed them immediately after feeding. To his surprise he observed large amounts of sugar in hepatic veins(Karamanou et al., 2016). He then analyzed liver tissue samples and in every liver he examined he found large quantities of glucose which were missing from other organs (Karamanou et al., 2016). He concluded that the liver was storing a water insoluble substance which is known today as glycogen which was converted into sugar and secreted into the blood (Karamanou et al., 2016). He concluded that it was an excess of this secretion that caused diabetes(Karamanou et al., 2016).

Bernard later was able to demonstrate the connection between the central nervous system and diabetes (Karamanou et al., 2016). Using a needle, he stimulated the fourth brain ventricle and produced temporary "artificial diabetes" which lasted less than one day (Karamanou et al., 2016. This procedure is named piqûre diabétique and linked for the first time glucose homeostasis and the brain to the pathogenesis of diabetes.

The work of Claude Bernard on glycogenic action of the liver highlighted the pathway of gluconeogenesis which supported the research on diabetes (Karamanou et al., 2016).

OSKAR MINKOWSKI AND JOSEPH VON MERING
An important turning point in the history of diabetes mellitus took place in 1889 after the experiments of Minkowski and von Mering.

In 1886, von Mering discovered that phlorizin, a derivative from glucose, could cause short-term excretion of large amounts of sugar in the urine (Karamanou et al., 2016. In 1889, while von Mering was working in Hoppe Seyler's Institute at the University of Strasbourg, Minkowski was an assistant at that time to the German leading scientist on diabetes Professor Bernard Naunyn and visited the Institute to look at some chemistry books of the library (Karamanou et al., 2016. They met accidentally and debated about the use of Lipanin then their conversation turned on whether the pancreas had a role in digestion and absorption of fats(Karamanou et al., 2016). Coming out of the discussion, the two decided the same evening to perform a pancreatectomy, the removal of the pancreas in a dog in the laboratory (Karamanou et al., 2016). The animal remained alive and was closely observed by Minkowski, as von Mering left urgently due to a family emergency (Karamanou et al., 2016). Soon after the operation, the dog developed polyuria (Karamanou et al., 2016. Minkowski examined the urine and found that it contained a high percentage of sugar. Initially, he was doubtful however, when he repeated the experiment in three more dogs, all of them developed glycosuria (Karamanou et al., 2016).

Moreover, Minkowski implanted a small portion of pancreas subcutaneously, in dogs without pancreases, and observed that hyperglycemia or high blood sugar levels was prevented until the implant was removed (Karamanou et al., 2016).

Minkowski and von Mering experiment demonstrated that the pancreas was a gland of internal secretion and that it played an important role for the maintenance of glucose homeostasis (Karamanou et al., 2016). They also set a foundation for Banting and Best to conduct their experiments.

FREDERICK BANTING, CHARLES BEST, JAMES BERTRAM COLLIP AND JOHN MACLEOD

In 1923, the Nobel Prize in Medicine was awarded to Frederick Banting and John MacLeod for the discovery of insulin. It was actually a story of success that provoked a great scientific conflict.

Frederick Banting was a young Canadian surgeon, who was admitted into the laboratory to study diabetes with Professor John Macleod, at the University of Toronto(Karamanou et al., 2016). In 1920, Moses Barron, published an article on "The relation of the islets of Langerhans to diabetes", which mentioned the further exploration of experiments of Minkowski and von Mering could lead to the discovery of the control of diabetes (Karamanou et al., 2016). Banting was influenced by this article (Karamanou et al., 2016).

In May 1921, Banting initiated a collaboration with Charles Best, a young medical student. Experimenting in dogs they initially ligate the pancreatic ducts, to reduce the exocrine region and almost ten weeks later they removed the dog's degenerated pancreas (Karamanou et al., 2016). They crushed the reduced pancreatic glands in a mortar and frozen in salt water. Then the mass was ground down and added to 100 mL of salt (Karamanou et al., 2016). Afterwards, they administer 5 mL of this extract intravenously to a dog without a pancreas(Karamanou et al., 2016). Within 2 hours, its blood sugar had decreased significantly. This experiment was repeated several times with other diabetic dogs, showing similar results (Karamanou et al., 2016).

At the end of 1921, the chemist James Collip joined the team and helped develop a better extraction and purification technique. The substance that was obtained was initially named by the team insletin but renamed later on by MacLeod to be insulin (Karamanou et al., 2016).

The next step was to test insulin in humans. So in 1922, insulin was administered to Leonard Thompson, a 14-year-old boy treated for diabetes in Toronto Hospital (Karamanou et al., 2016). After the introduction of Apollinaire Bouchardat's leading dietary treatment for diabetes, physicians told patients to eat as little as possible (Karamanou et al., 2016). Thomson at the time was also following a strict fasting diet suggested by Frederick Madison Allen (Karamanou et al., 2016). He was in critical condition. He received 15 mL of insulin through an injection in his buttock but he developed an infection at the injection site and became even more sick (Karamanou et al., 2016). Collip further improved the quality of insulin as 12 days later, Thompson received a second injection (Karamanou et al., 2016). The results proved to be excellent. His blood glucose from 520 mg/dL fell to 120 mg/dL, a normal blood glucose level in a period of 24 hours and urinary ketones disappeared (Karamanou et al., 2016). Thompson continued the insulin treatment and lived another 13 years (Karamanou et al., 2016).

The ground breaking work of Banting and Best saved millions of lives and gave diabetics a chance to live a normal life (Karamanou et al., 2016). Lilly Pharmaceutical Company worked with the two scientists and in 1923 introduced Iletin, the world's first commercially available insulin product (Karamanou et al., 2016). This took a turn when in 1923, the Nobel Committee decided to award Banting and MacLeod for insulin's discovery (Karamanou et al., 2016). Banting was upset by this news as he believed that he should share the prize with Best instead of MacLeod so he decided to share with Best his cash award as MacLeod shared his award with Collip (Karamanou et al., 2016).

CONCLUSION

For more than 3000 years, physicians have been on the quest to find the causal mechanisms and treatment of diabetes mellitus. However, an important progress has been made over the last 2 centuries thanks to the development of chemistry, physics and pharmacology. Over the next years scientists continued to make significant discoveries that allow for the development of novel treatments possible.

2

DISCOVERY OF DIABETES

By Eline El-Awad Gonzalez

An ailment presumed to be diabetes was first identified in manuscripts of Hesy-Ra, an Ancient Egyptian physician, dating back to 1552 BC. Among the oldest and most influential papyri of Ancient Egypt, the Ebers Papyrus enclosed the first medicinal insinuation to diabetes stating "… to eliminate urine which is too *asha*". The aforementioned term meaning 'ample' or 'excessive' could have referred to extreme urination caused by diabetes or a urinary tract infection. The Egyptians had prescribed combinations of natural remedies and rectal injections of honey, olive oil, sweet beer, sea salt, and wonderfruit seeds (Sanders, 2002). The Ebers Papyrus contained elixirs "to eliminate urine which is too plentiful" including:

A measuring glass filled with Water from the Bird pond, Elderberry, Fibres of the asit plant, Fresh Milk, Beer-Swill, Flower of the Cucumber, and Green Dates.

Egyptian medicinal practices shaped bordering cultures such as that of the Ancient Greeks. Following Galen's guidance, Aretaeus of Cappadocia—one of the most distinguished Greek physicians—attempted to treat diabetes but was unable to achieve a satisfactory outcome. Hippocrates publicized the conception of preventative treatment, emphasizing the importance of lifestyle changes, exercise, and nutrition on wellness. Alternative approaches of therapeutics included wine, overindulgence of food for fluid-loss compensation, famine diets, and much more. Greco-Roman accounts did not however describe a definitive record of diabetic conditions. Roman writer Rufus of Ephesus wrote about diagnoses founded on "incessant thirst" and urgent urinary excretion following consumption of beverages. Byzantine writer Oribasius echoed the work of Galen and Rufus, describing the symptoms as a *diarrhea urinosa* (Christopoulou-Aletra & Papavramidou, 2008).

The medicinal Hindu system, Ayurveda, was developed during the Vedic period ranging from 1750-500 BCE. Excessive urine described in Sanskrit texts dating to the fifth century was affiliated with a sweet taste. Ayurveda offered regime therapies for the ailment with references to the distinguished 'honey urine'. It considered the functionality of individuals in the framework of achieving a counterbalance between living forces and *doshas* possessed by each spirit (Gordon et al., 2019). The *doshas* of the body consist of *vata*—air, *pitta*—fire, and *kapha*—earth & water, while the *doshas* of the mind include *satva, rajas,* and *tamas.* Establishing an equilibrium amongst the *doshas* anatomically and operationally, appears as good health; however, an imbalance leads to illnesses. The Ancient Indians (circa 400–500 AD) were conscious of this urinary malady and had developed a primitive diagnostic test for *madhumeha.* Sushruta, an Indian healer, had primordially detected the affinity of ants and flies to the urination of those afflicted by this condition. The Hindu physicians Charaka, Susruta, and Vaghbata, described polyuria and glycosuria thousands of years prior to European depictions of an increasingly encyclopedic conceptualization of diabetes.

The Yellow Emperor's Classic of Internal Medicine, or Huáng Dì Nèijīng (黃帝 口 經), was among the most elemental ancient texts in Chinese treatment and a primary apostle of the Daoist philosophy harping the significance of living in harmony with *Tao.* Similarly, in the Treatise on Cold Damage and Miscellaneous Diseases, or Shānghán Zábìng Lùn (傷寒雜病 論), Zhang Zhongjing recorded nine subsections and nine herbal antidotes for diabetes-like symptoms. The text outlined an upper (lungs), middle (stomach), and lower (kidneys) association with the excessive urine. During the mid-1100s, Liu Wansu expanded on Zhang's commentary stating an attribution of *kidney-yin* deficiency to glycosuria. Dating back to the Warring States (475–221 BC) and Western Han dynasty (206 BC-8 AD), texts recorded symptoms of xiāo kě (消渴) as "three increases [excess] and one decrease [loss]". The three excesses included thirst (polydipsia), hunger (polyphagia), urine (polyuria) and one decrease in weight. These pillars progressively corresponded to modern-day standard stages of diabetes. Sui and Tang dynasties greatly augmented the diagnosis and treatment of xiāo kě. Subsequently, Zeng Liyan constructed the detection of contemporary diabetes mellitus through quantification of sugar in urine. Such appraisal was echoed by practitioners of the generations that followed. Particularly, the Wàitái Mìyào (外臺秘要) texts included Wang Tao's exhaustive record of sweet urine and a synopsis of historical diabetology archives preceding the Tang regime. In 600 AD, Sun Simiao continued to formulate treatment advances for prevention, ordinance, and tending of the ill. The herb assortment broadened from *Eupatorium fortunei* to dozens by Zhang and hundreds by Sun. *E. fortunei* has long been applied to nurse nausea and poor appetite while being prescribed in Chinese medicine as a diuretic and detoxifying agent (Tan et al., 2015).

Under the Abbasid Caliphate during the Islamic Golden Age, Muslim practitioners preserved, classified, and cultivated primeval remedial records from the Eurasian continent. Their manuscripts paved the groundwork for future developments in European medicine following the acquisition of Islamic authors' work during the 12th century Renaissance. The Canon of Medicine—a comprehensive 1025 medicinal encyclopedia compiled by the court physicians of the caliphs of Baghdad, Avicenna—enclosed clinical characteristics of diabetes diagnosed by an aberrant appetite, prostration of sexual functions, and sweet tang of urine (Eknoyan & Nagy, 2005). Avicenna had suggested that the condition was caused by *the sweet waters of the Nile* and the conventional heat that propagates over the kidneys (Chapter

1). He further detailed diabetic tissue death (gangrene) and treated patients with an herbaceous combination of lupine (a short-lived, attractive legume with flower spikes), fenugreek (a medicinal herb containing golden seeds, green leaves, and white flowers), and zedoaria (a perennial herb native to South Asia bearing yellow flowers and a large edible rhizome). Notably, this treatment was reported to be effective in five distinct cases in Tunis. Recent studies have suggested that the yellow seeds of the fenugreek are able to stimulate insulin production in diabetics (Sanders, 2002).

During the Middle Ages ranging from the fifth to the late fifteenth century, physicians diagnosed patient conditions using uroscopy. By tasting a patient's urine, an elaborate urine flavour diagram describing the taste, smell, and outlook of the sampled urine is used for diagnoses. Such a diagnostic tool allowed for fairly accurate diagnosis of diabetes based on the discrete sweet taste of the patient's urine.

Fast forward to the 18th and 19th century, Johann Peter Frank distinguished diabetes mellitus and diabetes insipidus. In 1776, Matthew Dobson verified the presence of excess sugar-derivative in the urine (glycosuria) and blood (hyperglycemia) inducing the historically noted sweet taste. Dobson offered empirical evidence by heating half a gallon of urine till it was dry, thus leaving a white residue that "was granulated and broke easily between the fingers; it smelled sweet like brown sugar, neither could it be distinguished from sugar, except that the sweetness left a slight sense of coolness on the palate." By 1841, Karl Trommer would develop the first clinical test for glucose in the urine, employing acid hydrolysis. Alternatively, urine was boiled by clinicians for determination of glucose, and ketones were estimated by pellets that altered colour. Treatment for diabetes frequently consisted of alloys containing opium. However, a misguided remedy consisted of nourishment appertaining to overcompensation of lost nutrients. Patients were recommended to gorge excess quantities of sugars, eventually leading to early deaths.

Following this incidence, greater success was established as physicians noted the attainment of fasting in symptom improvements for patients with diabetes.

Throughout the Franco-Prussian War of the early 1870s, French practitioner Apollinaire Bouchardat developed personalized diets as a diabetic treatment. Such conclusion came from his observations of improved symptoms in diabetic patients caused by wartime rationing of food (McCoy, 2009). By the 20th century, Frederick Allen published "Total Dietary Regulations in the Treatment of Diabetes" presenting a treatment of rigorous diets and famishment therapy. This incited the early 1900s fad diets, which included the "oat-cure," "potato therapy," and "starvation diet." A 2019 crossover study did however reveal a significantly reduced daily insulin requirement (82.0 and 69.9 IU versus 112 IU) in patients consuming oatmeal for two days (Delgado et al., 2019).

Following the introduction of these diets, physicians rapidly reignited the promotion of starvation diets in combination with routinely exercise to tackle this ailment. However, despite efforts to control this condition through lifestyle changes, patients with diabetes inevitably died hastily. The beginning of the largest breakthrough that would later lead to the discovery of insulin as a diabetes treatment was in 1889. Researchers at the University of Strasbourg in France, Joseph von Mering and Oskar Minkowski, revealed that the expulsion of a canine pancreas may stimulate diabetes (McCoy, 2009). In 1910, Sir Edward Albert Sharpey-Schafer suggested that diabetes was induced upon the absence of a distinct synthetic originating from the pancreas. He referred to it as insulin— meaning island—on the grounds that the cells of the pancreas in the islets of Langerhans had released it (Weatherspoon, 2020).

Eleven years later, Frederick Banting and Charles Best reversed evoked diabetes in dogs by injecting them with extracts of the pancreatic islets of Langerhans of wildtype controls. Following the purification of the insulin hormone from cow pancreases, a novel diabetes treatment had been created. In 1922, 14-year-old Leonard Thompson was administered the first insulin injection for diabetes treatment, allowing him to live for another 13 years. Banting and MacLeod later went on to receive a Nobel Prize in Physiology or Medicine, followed by announcing free access to insulin for diabetes patients. Additionally, the Oral Glucose Tolerance Test was introduced to measure the body's response to glucose as a tool for type 2 diabetes screening. This test, recommended by the World Health Organization, measures the speed of which glucose is vindicated from the blood following its introduction. Many variations have since emerged varying in dosage, sample intervals, and administration routes.

In the same year, Emil Werner and James Bell had first described metformin in scientific literature, as the outcome of N,N-dimethylguanidine synthesis (Werner, 1922). This was followed in 1929 by Slotta and Tschesche detecting metformin's sugar-reduction activity in rabbits, appointing it to be the most potent biguanide chemical that they had investigated. By 1972, Canada had approved metformin for type 2 diabetes followed by the U.S. Food and Drug Administration (FDA) approval in 1994. Metformin was soon overshadowed by alternative guanidine synthetics such as synthalins and ultimately, insulin.

In the early 1960s, urine strips were developed for sugar detection and simplification of blood sugar managing systems. Additionally, single-use syringe improvements permitted accelerated insulin therapy alternatives. By the late end of the decade, large portable glucose meters were produced to manage diabetes. They allowed for monitoring of blood sugar extents from anywhere while providing precise outcomes. Several forms of the blood glucose meters are accessible for at-home employment. A plethora of models had developed from basic-use reading only glucose levels in the blood to more advanced editions offering memory and storage features (Basina, 2018).

By the mid-1970s, insulin pumps were refined to mimic the normal discharge of insulin by the human body. These small, computerized devices are worn outside the body and deliver insulin continuously through a catheter. The pump would soon replace the need of daily insulin injections for diabetic patients. However, the risk of diabetic ketoacidosis was often overlooked as the body commences the breakdown of fat much too rapidly. Furthermore, pump failure caused by a displaced catheter or an automated accident, lead to colossal halts in production.

By 1982, the American biotech company, Genentech, cultivated the first biosynthetic human insulin—Humulin. The synthetic was licensed to Eli Lilly and became the first vendible product produced by recombinant DNA technology. Recombinant pharmaceuticals are developed through insertion of distinct genes from one species into a host microbe not containing it naturally—in this case bacteria. The organism can then be used as a 'living factory' to generate the desired outcome as the inserted genes code for the chosen product. In the present case, the gene encrypting human insulin is introduced into *Escherichia coli*. The large quantities of insulin produced by the bacteria is then collected and employed as the active Humulin component. The purified product is allocated to pharmacology stores for the use by diabetic patients. Primitively, this advancement was disregarded by the medical community as a non-scientifically significant improvement. Nevertheless, by the mid-1990s, the emergence of insulin derivatives had improved excretion, metabolism, distribution, and absorption characteristics immensely based on this initial biotechnological occurrence.

Since the development of metformin and insulin, other prominent revelations included (Patlak, 2002):

- 1942: The first sulfonylureas able to increase the release of insulin from the pancreas were classified. Sulfonylureas are medicinal organic compounds specifically targeting management of diabetes mellitus type 2.

- 1946: The Neutral Protamine Hagedorn insulin was developed by Novo Nordisk, having the asset of being combined with insulin resulting in its long-lasting action.
- 1950s: Biguanides uses for Type 2 diabetes able to reduce glucose production following food digestion were introduced. Additionally, Rosalyn Yalow and Solomon Berson discovered the radioimmuno-assay for insulin.
- 1955: Frederick Sanger determined the amino acid sequence of insulin, the first of any protein. He was later awarded a Nobel Prize for his work.
- 1969: The 3-dimensional structure of insulin was determined. Single crystal x-ray analysis was employed for its determination as a stable 36,000 molecular weight hexamer in the presence of zinc ions.
- 1988: The collection of symptoms (now called metabolic syndrome) was identified by Gerald Reaven. The cluster of conditions occurring synonymously, increase risk of heart diseases, type 2 diabetes, and strokes.
- 1996: Thiazolidinediones were introduced as an effective insulin sensitizer and approved by the FDA. These oral hypoglycemics function contrastingly with biguanides such as metformin.

In 2005, marketed as Byetta, exenatide was approved to treat type 2 diabetes in patients unable to reduce their blood sugar (Pollack, 2005). This injectable drug emulates the influence of the hormones produced and released by the intestine in response to digestion. These hormones, named incretins, increase insulin secretions, slow glucose absorption, and diminish glucagon activity. A recent 2019 review of Ayurvedic therapies for type 2 diabetes treatment revealed promising intervention potentials. Although Ayurvedic customs primarily highlight the use of herbs, supplemental factors such as weight management are accentuated as well. The objective of such practices is to convey hemoglobin A1c degrees into restorative ranges. A case study in Bangalore details a decrease in HbA1c levels from 11.2 to 5.7 in nine months following the prelude of Ayurvedic practices.

Currently, the A1c testing tools are used to mediate blood sugar levels over long periods of time rather than a one-point. While blood tests are able to diagnose diabetes, this test served to indicate the management therapy for diabetes. Although HbA1c remains the standardized measurement for diabetes, other testing methods are also employed, such as Fasting Plasma Glucose, Random Plasma Glucose, and Capillary Glucose.

Diagnostic and management tools for diabetes are continuing to improve following the rapid evolution of existing technologies. Long-term control of blood glucose continues to be a crucial landmark needed for improved management of diabetes symptoms. Awareness of such necessities is necessary for effective patient guidance and protection. The advantages of offering a broad spectrum of treatment possibilities include reduction of comorbidities with contiguous utility of supplement and spiritual practices. Although rudimental, ancient societies' treatments have proven to have a precise impact on diabetes treatment, particularly diet-based practices. As such, the following chapter will explore the impact of research pertaining to the efficacy of the major discoveries and how findings have evolved over time.

3

IMPACT OF RESEARCH

By Ayah Nour Nehdi

Diabetes is by no means a recent disease. Some of the earliest descriptions of diabetes date more than 3,500 years ago. One of the first documentations which mention diabetes traces back to a collection of ancient egyption medical manuscripts, known by the name of the Ebers Papyrus, that date back to around 1553 years before Christ (BC) (Vecchio et al., 2018). The greek physician, Aretaeus of Cappadocia (A.D. 30-90), describes what would most likely be type 1 diabetes (T1D) as a "melting down of flesh into urine, with short survival" (Zinman et al., 2017). While historical understandings of diabetes are quite limited, in some ways, they allowed past civilizations to respond creatively to the disease. Like described in chapter 2, diabetes was diagnosed and tested for with ants and urine in Ancient India (Weatherspoon, 2020). If ants were attracted to an individual's urine, this was an indication of a high glucose content, and thus diabetes (Weatherspoon, 2020). However, the treatments associated with historical definitions for diabetes were not nearly as effective as current therapies. Some of the treatments preferred by the Ancient Egyptians include elixirs made from herbs, plants, and natural substances(3). On the other hand, the ancient Greeks recommended treating diabetes with exercise, such as horseback riding, because it was believed that such activity would inhibit excess urination (Weatherspoon, 2020). Certainly, these ancient treatment suggestions were ineffective since they ultimately failed to address the pathology of diabetes, a disease characterised by insulin resistance (T2D) or a lack of insulin producing cells (T1D). Yet, it was not until the 1920s that effective pharmacological therapies were developed to manage diabetes.

Diabetes mellitus was a fatal maladie before the discovery of insulin in 1922 (Rockefeller University, 2021). At the beginning of the 20th century, the majority of scientists recognized diabetes as a sort of disorder where the patient expresses high blood glucose levels (Rockefeller University, 2021). Physicians and medical research were aware that some sort of substance from the pancreas was involved in regulating insulin, but weren't sure exactly what it was until 1910 when insulin was discovered, but not yet isolated (Rockefeller University, 2021). Fredrick M. Allen (1879 – 1964) was the first to characterize diabetes as a global metabolic disorder (Rockefeller University, 2021). Before 1922 and the isolation of insulin, Allen was also the first to develop a therapy for diabetes that managed to extend a diabetic's life, but was not curative (Rockefeller University, 2021). This therapy, coined as the "starvation therapy", consisted of a calorie-restricted diet. The aim of undernourishment in diabetic patients was to provide the minimum amount of carbohydrates necessary to sustain life and thus reduce glycosuria, the presence of glucose in urine (Rockefeller University, 2021). Allen's theory supporting undernourishment therapies derives from his previous research and experiment where he performs partial pancreatectomies on hundreds of dogs, cats, and other animals to induce them with diabetes (Mazur, 2011). This experiment allowed Allen to investigate diabetic metabolism via the direct quantification of the animal's food intake and urine-glucose levels and ultimately determine that calorie-restriction decreased glycosuria (Rockefeller University, 2021). Allen and Elliot P. Joslin, both prominent diabetes specialists in the United States, promoted such "starvation diets", which consisted of repeated fasting and prolonged undernourishment, as the most advanced and effective therapy for T1D (Mazur, 2011). However, such a treatment was not curative and could only provide relief of T1D symptoms with a possible extension of life (Mazur, 2011). Today, there is evidence and research supporting that fasting reduces glucose levels in diabetics. However, while calorie restricted diets can be of benefit for diabetic patients who are overweight, prolonged calorie-restriction introduces many health risks, such as a compromised immunity, stunted growth in children, and potential death by starvation (Mazur, 2011). Ultimately, Allen's fasting and under-nourishment therapy for diabetes was "unpleasant, difficult to maintain, and enervating" - all of which exceeded the potential benefits for many patients (Mazur, 2011). The therapy was unattainable for many patients since they sooner or later withdrew from the program (Mazur, 2011). Allen and Jos-lin disliked prescribing the therapy, especially to children. Joslin once com-mented that "it was no fun to starve a child to let him live" (Mazur, 2011). Other physicians criticized that starvation and a poor quality of life associat-ed with the starvation therapy was too high of a price to pay for a rather small extension of life (Mazur, 2011).

Elizabeth Hughes, the daughter of Charles Evans Highes who was the Gov-ernor of New York and a Republican candidate for the presidency in 1916, was Allen's most famous recognized patient (Mazur, 2011). She was diag-nosed with diabetes in 1919 at age 11 (Mazur, 2011). Initially, Allen pre-scribed Elizabeth with a week of fasting which was followed with a daily count of 500 calories along with one fast day per week (Mazur, 2011). Once Elizabeth was freed of glycosuria, her calorie count was increased to 1250 with one fast day per week (Mazur, 2011). At the time of her diagnosis, Eliz-abeth weighed 75 pounds but she continued to deteriorate to a staggering 45 pounds due to the severe fasting therapy (Mazur, 2011). While it is difficult to gage how long Elizabth's life was extended because of Allen's restricted diet regimen, she was still able to survive three more years past her diagno-sis, allowing her to be included in an insulin clinical trial conducted by the Canadian Doctor Fredrick Banting (who discovered insulin). The clinical tri-al proved successful as Elizabeth regained weight and went on to eventually graduate from Barnard College, marry, have three children, and be involved throughout her life in civic affairs – all the while on insulin. While some of Allen's patients were able to survive past 1922 and receive insulin treat-ments, many were not so lucky. Allen and other physicians treated about 100 diabetic patients between 1914 – 1917 at the Rockefeller Hospital using starvation diets (Mazur, 2011). Of those patients, one was a seven-year-old boy first seen in 1915 (Mazur, 2011). On admission, the boy weighed 40 pound and was discharged at 37 pounds with a prescribed 700 calorie diet along with one fast day per week (Mazur, 2011). Unfortunately, the boy died two months later, details of which were not obtained. By 1917, 43% of Allen's 76 diabetic patients died while on diet restrictive therapies (Mazur, 2011). While mortality associated with the starvation therapy was high, Joslin notes that prolonged fasting was involved in the reduction of coma and thus consequent deaths associated with diabetes. However, since Allen's therapies and patient cases lack an equivalent and proper control group, it cannot be certain whether undernourished patients fared a better fate than diabetics who maintained healthy diets (Mazur, 2011). At the turn of the 20th century, physicians were only able to measure blood sugar levels in urine and no other effective treatments and so it comes at no surprise why Allen's restricted calorie diet became quickly adopted – diabetics had to choose between starvation or death (Mazur, 2011). Allen's restricted calorie diet and prolonged fasting developed in 1913 was successful in mitigating glycosuria, acidosis, coma, and other horrible outcomes of the disease (Mazur, 2011). Nonetheless, undernourishment was not curative and unsustainable for the long term.

Undoubtedly, the discovery of insulin proved to be a milestone in medical history when diabetes was no longer thought to be a death sentence. By 1910, Insulin had been identified as a substance capable of mitigating glucose metabolism and was secreted by pancreatic islands (Vecchio et al., 2018). Many scientists encountered a plethora of difficulties isolating insulin, especially in the extraction of the Lengerhans islands from the rest of the pancreatic exocrine tissue (Vecchio et al., 2018). However, by utilizing a specific technique, Banting was able to obtain pancreatic islets in a pure state by closing pancreatic ducts allowing for the degeneration of the pancreatic exocrine tissue and thus the isolation of an insulin extract. Isolated insulin allowed Banting to control glucose levels in diabetic animals and then in diabetic humans. Leonard Thomspon, as discussed in the previous chapter, was the first patient to ever receive an insulin injection which proved successful as it led to normalization of blood glucose levels, glycosuria, ketonuria (the presence of ketone bodies in urine). It is important to note that banting's insulin was short and rapid acting, lasting a maximum of 6 hours (Vecchio et al., 2018). There was a significant need for extended-action insulin since patients otherwise would have to receive multiple doses of short-acting insulin throughout the day and during night-time to regulate blood sugar levels (White, 2014). Diabetic children who did not receive consistent insulin injections, especially throughout their sleep, were at a significant risk for reduction in growth and diabetic dwarfism syndrome (also known as Mauriac's syndrome). The advent of crystallization of insulin was a major advancement that opened the door to many insulin modifications, especially those with different time-action profiles (White, 2014). Yet it wasn't until 1936 when the first extended-action insulin, PZI (protamine zinc insulin), was commercially available (White, 2014). Diabetic patients today use both short acting and long acting insulin, the first for controlling glucose spikes throughout the day, and the latter to provide background regulation. Over the course of the 20th century, a myriad of formulation changes were made to improve insulin quality, including the production of human insulin analogs which provide more effective treatment (White, 2014). The evolution of insulin therapies has provided diabetic individuals today with significantly higher life expectancy (White, 2014). It is reported that the life expectancy for patients with type 1 diabetes (aged 20 – 24) is around 11 – 14 years higher than individuals without diabetes (White, 2014). Whereas data reported in 1975 suggests a 27 year lower life expectancy for individuals with type 1 diabetes (White, 2014).

Not only has the development of insulin and its pharmacological enhancement drastically impacted diabetic patients, but so have improvements in insulin delivery systems and other related technology. Insulin delivery is a critical area of research and improvement due to the fact that type 1 diabetics usually require at least 2, but more often 3 or more daily injections (Al-Tabakha and Arida, 2008). From the perspective of an entire diabetic patient's lifetime, it is estimated that 60,000 units of insulin would be injected to sustain life (Al-Tabakha and Arida, 2008). Thus, not only producing effective delivery methods is of utmost importance, but so is producing convenient and pleasant insulin delivery experiences. Studies suggest that once non-invasive insulin deliveries are available, 2.5% of the total diabetic population will use them with projected rapid increase by time (Al-Tabakha and Arida, 2008). Subcutaneous injections are the most traditional and predictable insulin delivery system. However, subcutaneous administration tends to be rather painful and thus a deterrent to patient compliance since many require multiple doses per day. Many changes have been adapted to decrease diabetic patient suffering and any other inconvenience, some include: supersonic injectors, infusion pumps, sharp needles and pens. Traditional syringes were initially large, heavy, and bore a large needle (Al-Tabakha and Arida, 2008). Additionally, syringe administration in social settings, such as public spaces, classroom, and workplace, may be considered as drawbacks due to its complicated administration, lack of discretion, and inconvenience (Al-Tabakha and Arida, 2008). It wasn't until 1970 when insulin pumps and 1985 when insulin pens were first introduced. Numerous studies confirm that insulin pens provide greater accuracy, ease of use, patient satisfaction, quality of life, and consistent adherence (Selam, 2010). Moreover, pens have become widely accepted, especially in Europe amongst the elderly and the adolescents, generally due to their greater comfort and simple administration (Al-Tabakha and Arida, 2008). Database analysis by the United States also indicates that compared to insulin delivery via vial and syringe, improved adherence associated with the use of insulin pens could potentially curtail diabetes care costs (Selam, 2010). On the other hand, the advent of Insulin infusion pumps allowed for greater flexibility and personalization in insulin delivery for patients based on their needs (Al-Tabakha and Arida, 2008). Insulin pumps have significantly impacted diabetic patients allowing them to successfully control their blood glucose levels and hence, their quality of life. Several pharmaceutical companies are currently researching and developing new methods of insulin delivery, including pills, patches, inhalers, and mouth sprays (Al-Tabakha and Arida, 2008). Some insulin inhalers have already been approved for marketing. In 2006, the FDA approved Exubera®, an inhalable insulin developed between Sanofi-Aventis and Pfizer, for use in patients over 18 with T1D and T2D (Al-Tabakha and Arida, 2008). Exubera® has also been approved by the European Commision (EC) for T1D and T2D that same year (Al-Tabakha and Arida, 2008). Studies also suggest that converting from subcutaneous insulin injections to inhalations (similar to asthma inhalers) will result in higher adherence to therapy and thus improved glycemic control (Al-Tabakha and Arida, 2008). However, Exubera® was pulled from the market in late 2007 due to its high cost, low sales, and other dosage issues (Al-Tabakha and Arida, 2008).

Alongside technological improvements and research on insulin delivery methods, advancements in the technological monitoring of blood glucose levels was a crucial factor that allowed greater personalization in insulin treatments. Specifically, digital blood glucose monitors, which were developed in the 1980s for portable use, allowed diabetic patients to consistently measure their blood-glucose levels and thus, have better control over their insulin treatments (Weatherspoon, 2020). Glucose monitoring technologies are necessary in controlling glycemic levels for patients with T1D (Al Hayek et al., 2020). Efficient management can significantly reduce medical risks such as microvascular and macrovascular complications (Al Hayek et al., 2020). Yet, a significant proportion of diabetic patients, especially young adolescents, fail to adhere to strict glucose self monitoring via finger pricking due to their fear needles(11). With scientific technologies improving exponentially, there have been many breakthroughs in developing alternative self glucose-monitoring systems that can provide greater acceptability and ease. In particular, the FDA approved of the Freestyle Libre Pro in 2016, a flash glucose monitoring system (FGM) where a disposable sensor is placed on the patient for up to 14 days (Blum, 2018). The technology allows diabetic patients to moniter their glucose levels more frequently and effortless since results from the FGM are sent to a personalized portable device and can be viewed any time simply by scanning the sensor (Al Hayek et al., 2020). In a cross-sectional study, adolescents between the ages 13 and 19 with T1D were placed on the Freestyle Libre Pro FGM system for at least 6 months – after which they responded to a questionnaire in regards to their acceptability of the technology. Of the study population, 95.5% agreed that the application of the sensor caused less pain than typical finger pricking; 89.6% reflected that the sensor did not disturb their daily activities. In comparison to freestyle techniques for glucose monitering, 86.6% reported that the FGM was les stressful, 95.5% suggested that the FGM was easier, 83.6% agreed that the system was more discrete and more comfortable. Without a shadow of doubt, these findings demonstrate immense acceptibility to FGM systems and the FreeStyle Libre Pro technology. For diabetics, measuring blood glucose levels is imperative in determining how much insulin they require and the effectiveness of their treatment. Developments in insulin delivery technologies like the FreeStyle Libre Pro help diabetic patients manage their condition themselves and with much greater ease, thereby providing more control and confidence in their ability to manage their blood-glucose levels. Diabetes research and development, especially during the 20th century, is a significant contributor to the modern day standard of diabetes treatments and understandings. The current armamentarium of technologies, types of insulin, and other drug therapies have resulted in a drastic decrease in morbidity and mortality compared to that of the pre-insulin era. The most prominent and accepted diabetes treatment before the advent of insulin, such as Allen's starvation therapies, involved significant health risks and were not nearly as effective as insulin. The discovery of insulin and its further modification, such as its crystallization which allowed for long-acting insulin coverage, has drastically enhanced a diabetic patient's ability to control their blood-glucose levels. Blood-glucose monitoring and insulin delivering technologies and their evolution throughout the past century has allowed diabetics to better adhere to their insulin therapy more consistently and with greater ease. Without a doubt, diabetic patients today have a greater outlook on life due to a significantly higher lifespan due to medical research and innovation. The relevancy of diabetic research has only been amplified due statistics indicating the proportion of individuals with diabetes has only increased over the past few decades (Honeycutt et al., 2001). If diabetes trends and statistics in the United States continue to project as studies predict, the number of Americans diagnosed with diabetes will from 11 million in 2000 to 29 million in 2050, an increase of 165% (Honeycutt et al., 2001). Hence, there continues to be an exponential demand for diabetic research and improvements due to increasing predicted prevalence of the disease. While the medical breakthroughs throughout the past century have provided individuals diagnosed with diabetes a significant extension of life and potential for normalcy, they are not curative. Ultimately, if research and development of diabetes treatments advances as studies project, therapeutic options today will presumably be considered arcane in the future, especially when curative treatments are finally available.

4
THE IMPORTANCE OF RESEARCH

By Susie Woo

The importance of diabetes research has dramatically become more prevalent over the course of the 20th to 21st century. According to the World Health Organization, the diabetic population has grown from 108 million in 1980 to 422 million in 2014 across the globe (World Health Organization, 2021). In 2017, approximately 462 million individuals were reported to have type 2 diabetes, which represents 6.28% of the world's population at the time. The majority of people who were diagnosed with type 2 diabetes were above the age of 50, whereby 15% of individuals were 50-69 years old, and 22% were 70+ years old. The prevalence rate of type 2 diabetes was at 6059 cases per 100,000 in 2017, however, it is projected to increase to 7079 individuals per 100,000 by the year 2030 (Khan et al., 2019). Furthermore, the International Diabetes Federation (IDF) predicts that the number of cases of diabetes will increase to 642 million by the year 2040 (Einarson, Acs, Ludwig, & Panton, 2018). In Canada, it is estimated that over 2 million Canadians have been diagnosed with diabetes in which 90 to 95% of cases are type 2 diabetes (Diabetes Canada, 2019). Although this represents a 1 in 16 prevalence rate, when including individuals with prediabetes, this rises to a 1 in 3 chance (Diabetes Canada, 2019; The Conference Board of Canada). However, these statistics do not even compare to largely populated countries that are placed at the top in terms of having the greatest total number of individuals with type 2 diabetes. For instance, China was reported to have 88.5 million individuals, India with 65.9 million individuals, and the United States having 28.9 million individuals as of 2017 (Khan et al., 2019).

Due to the sheer amount of people suffering from chronic type 2 diabetes, there have been deep concerns on mortality rates across the globe. In a systematic literature review on cardiovascular disease in people with type 2 diabetes, it was evident that about half of the deaths were related to cardiovascular disease (50.3%). Across eight studies with a total of more than 3 million patients with type 2 diabetes, the average death rate was 9.9%. In fact, the major contributors were coronary artery disease (reported as coronary heart disease or ischemic heart disease), which was responsible for 29.7% of cases, and cerebrovascular disease (i.e. strokes), which was responsible for 11.0% of cases (Einarson, Acs, Ludwig, & Panton, 2018). Globally, diabetes is one of the top 10 causes of death, and has a 2 to 3 fold risk of all-cause mortality. Furthermore, the incidence of diabetes has increased by 102.9% whereas the global prevalence has increased by 129.7% from 1990 to 2017 (Lin et al., 2020). The overwhelming impact of type 2 diabetes across the world and the rate at which this disease accumulates signifies a danger to the health of individuals of all ages and backgrounds. Although this disease has its own toll on the body, it is actually the comorbidities that may develop with chronic diabetes that expands the severity of the diabetic crisis. Many of the comorbidities are also noncommunicable diseases (NCDs), which are chronic, noninfectious health conditions that cannot be spread to other people (Marcin, 2018). According to the World Health Organization, NCDs kill 41 million people each year which is equivalent to 71% of all deaths worldwide. Furthermore, cardiovascular diseases account for most NCD-related deaths at 17.9 million people annually, followed by cancers (9.3 million), respiratory diseases (4.1 million), and diabetes at 1.5 million people. Overall, these four diseases account for over 80% of all premature NCD deaths, and severely reduces quality of life and increases the incidence of disabilities (World Health Organization, 2021). Hence, it is evident that diabetes, especially type 2 diabetes, is a current, impeding concern that affects millions of people across multiple nations, developed or otherwise. Therefore, research on the underlying causes of diabetes is crucial to tackling this worldwide crisis and establishing proper policies and guidelines to support the overall health of the public.

There are several risk factors that are involved in the development of type 2 diabetes, and other comorbidities, and are a result of both genetics and the environment. Majority of the risk factors are based on influences from the environment and are modifiable when taking the right precautions. It is also important to note that type 1 diabetes has been speculated to have similar risk factors as type 2, however, the actual cause of type 1 is still unclear (NIH). On the other hand, type 2 diabetes has been widely studied in several aspects of research and has demonstrated to be a preventable disease just by changing unhealthy habits. It has been hypothesized that the most significant environmental risk factors for type 2 diabetes are dietary choices, smoking and physical inactivity, which leads to an increase in adverse health conditions such as obesity, hypertension, prediabetes, and increased blood lipid levels (Dendup, Feng, Clingan, & Astell-Burt, 2018).

Due to the exponential improvement in the manufacturing industry, food has become highly accessible for cheap prices. This is prominent in fast food restaurants, processed snacks, and sugary drinks. Highly processed foods have become abundant in calories, saturated and trans fats, sodium, simple carbohydrates, and added sugar. Fast foods are known to serve energy dense foods in large portion sizes (Mandal, 2019). Recent evidence has suggested a relationship between soft drink intake with obesity and diabetes. This is largely attributed to high fructose corn syrup intake which raises blood glucose levels to a dangerous level, and can be a dangerous promoter of insulin resistance, and thus, type 2 diabetes (Sami et al., 2017). According to the American Heart Association (AHA), the average US citizen consumes 17 teaspoons (71.14 grams) of added sugar every day, even though daily recommendations are only half of that amount for men, and a third for women and children. In terms of calories, men are recommended to consume 150 calories in added sugar per day, while women and children should consume about 100 calories. In comparison with highly marketed sweets and soft drinks, a can of coke (12 oz) contains 140 calories from sugar alone (Miranda). Even if an individual were to consume only one or more sodas per day, their risk for metabolic syndrome increases by 36% and their risk for type 2 diabetes increases by 67%, compared to someone who does not drink at all (Cleveland Clinic, 2020). In addition, high intake of red meat, sweets, and fried foods also increased the risk for type 2 diabetes. It is important to keep in mind that this risk is not necessarily due to a high consumption of kinds of foods, but rather the low nutritional quality in balancing overall health. Conversely, extensive research has revealed a negative correlation between consumption of fruits and vegetables with type 2 diabetes. The increased intake of essential nutrients, fibers and antioxidants infers a protective effect against chronic hyperglycemia and other comorbidities. This knowledge and awareness of the potential benefits of a balanced diet is key to reducing excessive calorie consumption and achieving better eating habits (Sami et al., 2017).

Research on the impact of diabetes and dietary changes has brought forth a new hope for diabetics in reducing harmful cardiovascular repercussions. Nevertheless, it is equally as important to maintain an individual's activity levels, such that the net caloric intake does not exceed to the point of high blood glucose levels, fat gain and morbid obesity. Thus, one of the risk factors that need to be taken into consideration, is the unprecedented levels of physical inactivity. On a global scale, 23% of adults and 81% of adolescents (aged 11 to 17 years) do not meet the World Health Organization recommendations for physical activity for health, which is said to be at least 150 minutes of moderate-to-vigorous activity per week or 30 minutes of daily equivalent activity (World Health Organization, 2021; Panahi & Tremblay, 2018). For adults in developed nations such as the Americas and Europe, physical inactivity is at its highest and can be attributed to the increased urbanization, changes in transportation, and use of technology (World Health Organization). In fact, the average American spends about 55% of their waking time being sedentary - any waking behaviour that requires low energy expenditure (≤ 1.5 MET) - and is comparable to Europeans who are estimated to spend about 40% of their leisure time watching television (Panahi & Tremblay, 2018). Furthermore, there is a positive, non-linear correlation between total sedentary behaviour and all-cause mortality when adjusted for physical activity. This indicates that lower levels of exposure to television are associated with smaller levels of sedentary-related risks. But when leisure behaviour exceeds 8 hours a day, mortality risk increases exponentially. Similarly, there is a positive, linear association between type 2 diabetes and total sedentary behaviour, where there is a 29% incidence that is estimated to be related to watching television (Patterson et al., 2018). In addition, a meta-analysis of six studies found a relationship between daily sitting time and all-cause mortality in which there was a 34% higher mortality risk for adults sitting 10 hours a day even after taking physical activity into account (Panahi & Tremblay, 2018). Age is also a significant risk factor in sedentary behaviour and is a critical issue in today's world. Approximately 3.2 million deaths per year are suggested to be caused by inactivity and the prevalence of sedentary behaviour is apparent in industrialized countries where chronic health conditions, such as hyperglycemia, are increasing and physical activity is declining. Older adults especially, have been suffering from declines in muscle mass, balance, strength and endurance. The barrier of their physiology due to age affects their ability to participate in exercise in general. However, research has shown that older adults who engage regularly in physical activity are able to improve functional independence and prevent chronic symptoms, whether they do aerobic or resistance exercise. This infers reduced risk of developing cardiovascular disease and type 2 diabetes in older adults who exercise at a moderate level. In a study done in older American women, it was shown that higher levels of physical activity resulted in 40% to 50% lower all-cause, cardiovascular disease and cancer mortality rates compared to women exercising at lower intensities (Taylor, 2013). The advantages that come with physical activity and exercise is significant for the diabetic population, as older adults are the most susceptible in being diagnosed with diabetes. More than two-thirds of those with diabetes above the age of 60 have died from heart disease. This means that approximately one-third of diabetes-related deaths occur in people under the age of 60. Children in particular have also been observed to have a higher incidence of type 2 diabetes within the last decade. For instance, the prevalence of type 2 diabetes in children and adolescents was 1.54 per 100,000 at a minimum, and could be as 11.8 per 100,000 in 15 to 19 year olds. This can also be attributed to lower physical inactivity levels and increased prevalence of obesity (Panagiotopoulos, Hadjiyannakis, & Henderson, 2018).

Hence, research in diabetes, especially type 2 diabetes, has proven to show promise in preventative measures taken in diet and exercise by increasing awareness and knowledge of the modern epidemic. The impact of type 2 diabetes is widespread and plays a part in the progression of other health conditions such as chronic kidney disease, diabetic retinopathy, nervous and cardiovascular related diseases.

Diabetes research does not only ameliorate an individual's health but also has a grand influence on the economic aspect on a global scale. The cost of diabetes is at least 3.2 times greater than the average per capita healthcare expenditure, rising to 9.4 times in the presence of certain complications. In the United States in 2012, the estimated total cost of diagnosed diabetes was amounted to $245 billion dollars, where $176 billion comprised of direct medical costs, and $69 billion was attributed to lost productivity due to work-related absenteeism, reduced productivity at home, unemployment due to chronic disability and premature mortality. The values are projected to increase to $622.3 billion in the year 2030, including $472 billion in annual medical costs. The insurmountable burden due to type 2 diabetes on the national economy infers a prominent impact on the individual citizen as well. For men, lifetime costs can range from $54,700 for those diagnosed above the age of 65, and can go up to $124,700 for those diagnosed between 25 to 44 years. In women, the lifetime costs range from $56,600 for those diagnosed above the age of 65 years, and can reach up to $130,800 in patients between 25 to 44 years (Cannon et al., 2018). In Canada, there was a total estimate of $2.18 billion in health care costs, such as hospital care, physician care, and drugs for diabetes. By 2022, approximately 2.16 million new cases of diabetes will arise and will correspond to $15.36 billion in healthcare costs. One study examined the potential impact of two intervention scenarios that could help reduce the costs related to diabetes. The first intervention aimed at a 5% (average) weight loss in the population. This standard has been shown to have a positive impact on glycemic and cardiovascular health, while being a realistic goal for the majority of the public. Some large-scale environmental changes to make this goal achievable would include the implementation of more walkable areas and improved nutrition labelling. As a result, the 10-year predicted risk of developing diabetes would drop to 8.67% and the reduced number of new cases would save the country $2.03 billion compared to baseline projections. The second intervention that was analyzed specifically targets those in the highest-risk decile (i.e. those who have a 10-year risk of developing diabetes ≥ 22.6%). The goal of this scenario is to achieve a 30% reduction in their risk for diabetes and may consist of individualized lifestyle intervention programs or follow a pharmaceutical approach that has been proven to be effective in randomized trials. This scenario would reduce the overall risk of diabetes development by 9.02% and would save the health sector $1.48 billion in direct healthcare costs (Anja & Laura, 2017).

Overall, epidemiology research on different risk factors and economic aspects that are a direct consequence of diabetes has been recognized to be important alarms for healthcare professionals and policy makers to develop better preventative programs on a global scale. By understanding the root cause of diabetes, it is possible to avert the negative health consequences and reduce the chance of developing common comorbidities. The most important take-away from diabetes research is that the majority of the risk factors are environmental, and thus modifiable. By promoting awareness of this epidemic, and actively participating in its prevention, it is possible to achieve a balanced and healthy lifestyle without having to experience these complications.

5

WHAT IS IT?

By Aleefa Devji

WHAT IS DIABETES MELLITUS?

In this section we will be exploring what exactly diabetes is, and how it affects the health of an individual. In doing so, we will first define diabetes and understand how it is related to gluconeogenesis and glycolysis before learning about the types of diabetes and how they relate to these processes.

Diabetes mellitus, commonly referred to as just diabetes, is a group of diseases which affect the metabolism of glucose in your body (Diabetes Canada, 2021). Glucose is a vital source of energy that our body both produces and subsequently consumes to carry out our daily activities. Glucose is also responsible for providing the energy that our tissues and muscles require.

In our bodies, the processes of producing and consuming glucose are called Gluconeogenesis and Glycogenolysis respectively. Gluconeogenesis is the metabolic pathway that humans use to produce glucose through the process of breaking down non-carbohydrates into glycogen, pyruvate or other intermediaries to glycogen (Wikipedia, 2021). From the intake of food, the human body uses proteins, or lipids to produce substrates for glycogenolysis - process of breaking down glycogen to produce glucose - which are glucogenic amino acids and glycerol. Although gluconeogenesis occurs as a result of consuming food, it can also occur during periods of fasting when fatty acids are broken down to make glucose molecules in order to maintain bodily functions. Generally, this process takes place in the liver, but in those individuals with diabetes the process of gluconeogenesis is increased with production in the kidney as well, and this therefore increases the overall glucose production in the body. Another process that is key to this glucose cycle is Glycogenesis, which is the process of synthesizing the glycogen from glucose substrates – this process allows for increased solubility of glycogen to ease the accessibility of breakdown into glucose molecules when needed for metabolism (Wikipedia, 2021). On the other hand, we also have glycogenolysis which is the breakdown for glycogen to provide the body with glucose when required. This process occurs in the liver, muscles, and the kidneys in order to maintain the body's blood glucose levels to prevent hypoglycemia or hyperglycemia from occurring. Hypoglycemia is commonly referred to as low blood sugar and hyperglycemia as high blood sugar.

When talking about diabetes the idea of maintaining blood glucose levels in the body is an important concept to understand. With the utilization of gluconeogenesis and glycogenolysis outlined above, the body is able to produce and store glucose from the food we intake but is also able to subsequently use these glycogen stores to prevent hypoglycemia between meals (Wikipedia, 2021). This process in the body is referred to a glucoregulation or the balance between glycogen stores in the body and the free glucose in the bloodstream.

Now that we have a basic understanding of how glucose is produced, stored, and utilized in the body we can explore why this is important to understanding diabetes. In the human pancreas, hormones are released to facilitate the metabolism of glucose when necessary – these hormones are primarily insulin and glucagon. Glucagon is produced by alpha cells in the islets of Langerhans of the pancreas and insulin is produced by the beta cells of the islets of Langerhans as these two hormones work closely together (*Glucagon*, 2018). In the body, glucagon is responsible for preventing blood glucose levels from dropping too low to a point of hypogly-cemia and in order to do this it stimulates the conversion of glycogen stores to glucose that will be released in the bloodstream through the process of glycogenolysis. Another way that glucagon produces glucose is through the production of glucose from amino acid molecules through gluconeogene-sis. In working alongside insulin to maintain blood glucose levels, gluca-gon is controlled by insulin which works to ensure that the glucose levels in the blood do not rise too high and cause hyperglycemia.

With our background in glucose regulation and the body processes in-volved, we can now begin to talk about diabetes more specifically. Diabetes is a chronic condition that affects how the human body turns food into ener-gy for metabolism or daily function and activity. To recap briefly, the food we eat is turned into glucose and released into your bloodstream but when our blood sugar rises too high insulin is signalled to be released from the pancreas. The insulin allows the body to keep that extra glucose in its cells for later use and converts it to energy stores (Healthline, 2018). For indi-viduals who have diabetes, insulin plays a large role because either their body is not making enough insulin or is not able to respond to the insulin as well as it should be to balance hyperglycemia. This can cause a greater deal of issues over time, as the prolonged increase in blood sugar can cause heart disease, vision loss or even kidney disease.

TYPES OF DIABETES

As mentioned above, diabetes in individuals can either be caused by the inability to make insulin or the inability of the body to respond to insulin appropriately but there are also other factors to consider in diagnosing the type of diabetes that an individual may have. Most commonly there are three types, Type 1, Type 2 and Gestational diabetes that occurs during pregnancy (Mayo Clinic, 2020).

Type 1 diabetes is thought to be an autoimmune disorder where the body halts the production of insulin by mistakenly attacking its own pancreatic cells where insulin is made (Diabetes Canada, 2021). This classification is genetic and is seen in approximately 5-10% of the population that has diabetes. The symptoms of this form are easily detectable and normally present in children, teens and young adults. Individuals with this condition are required to take insulin on a daily basis in order to survive and maintain their blood glucose levels within a healthy range as a result this classification is referred to as insulin-dependent diabetes. Unlike type 2 diabetes, there is no known way to prevent this form.

Type 2 diabetes is another classification where the body does not use the insulin effectively, or they are not able to produce enough of it (Diabetes Canada, 2021). As a result, individuals with this form of diabetes are also unable to regulate their blood glucose levels. Another term used to describe or explain this form of diabetes is insulin resistance, which develops over many years and as a result is more often diagnosed in adults – although it is more commonly being diagnosed in individuals early on in life. In individuals with diabetes, type 2 is prevalent in approximately 90% of cases. Some individuals may be at a genetic risk to developing type 2 diabetes if individuals in their family have the condition, but this form of diabetes is preventable and/or the onset can be delayed with healthy lifestyle choices, activity and maintaining a healthy weight. For some individuals these lifestyle choices are enough to manage this classification of diabetes, but for others medication or insulin therapy may also be required.

Gestational diabetes is another common form that develops in women who are pregnant but have otherwise never had diabetes (Diabetes Canada, 2021). Although gestational diabetes goes away after the baby is born, it can put the mother at an increased risk of type 2 diabetes later in life and also poses health risks for the baby such as obesity or even developing type 2 diabetes themselves later in life.

Another classification of diabetes is prediabetes which occurs in individuals with higher-than-normal blood glucose levels but are not yet high enough to classify as a diagnosis for type 2 diabetes (Diabetes Canada, 2021). For these individuals, the implementation of healthy lifestyle choices and healthy eating is crucial to prevent or delay the diagnosis of type 2 diabetes.

WHAT CAUSES DIABETES?

The effect that diabetes has on individuals is dependent on how elevated their blood glucose levels are. For those with prediabetes or type 2, symptoms may not be experienced but for those with type 1 the symptoms are often severe and present themselves with a rapid onset.

For individuals with type 2 diabetes, there are several factors that can have an effect on the development such as being overweight or lack of physical activity that leads to insulin resistance (Center for Disease Control and Prevention, 2019). Individuals with extra weight, especially around the belly area, are linked to insulin resistance and type 2 diabetes. Insulin resistance is a condition where the muscle, liver and fat cells are unable to utilize insulin effectively and as a result individuals require an increase in insulin production to maintain the demand but overtime the pancreas is unable to synthesize enough insulin and hyperglycemia occurs. Other factors that can affect the prevalence of type 2 diabetes in individuals are genes and family history, as with type 1 genes can play a role in the development of type 2 (National Institute of Diabetes and Digestive and Kidney Diseases, 2016). As well, some racial/ethnic backgrounds are more prone to the development of diabetes – some of these groups are including but not limited to African Americans, American Natives, Asian Americans, Hispanics/Latinos, Native Hawaiians, Pacific Islanders.

Individuals can be predisposed to diabetes as a result of their genetic makeup, or their lifestyle choices and eating habits can lead to the development of the chronic illness as a result of insulin resistance, but it can also be caused by genetic mutations, hormonal diseases along with damage to the pancreas (National Institute of Diabetes and Digestive and Kidney

Diseases, 2016). Monogenic diabetes appears in individuals with mutations in their genes and can be a result of even a single gene mutation. Although this genetic makeup can be passed down through families, it can also occur on its own. In most cases of monogenic diabetes, the mutation prevents the pancreas from making enough insulin to support glucose metabolism in the body (National Institute of Diabetes and Digestive and Kidney Diseases, 2016). Another illness that can result in the prevalence of diabetes is cystic fibrosis which produces thick mucus that scars the pancreas and therefore prevents it from producing the insulin. Pancreatitis, pancreatic cancer, or damage/trauma to the pancreas can also result in stunted insulin production (National Institute of Diabetes and Digestive and Kidney Diseases, 2016). Along with this, some hormonal disease such as Cushing's syndrome, Acromegaly, and Hyperthyroidism can result in the presence of diabetes because they produce too much of a certain hormone that consequently causes insulin resistance.

As we can see from the host of potential causes, diabetes can affect anyone and does not discriminate. Therefore, it is important to take precautions in maintaining a healthy lifestyle, healthy eating choices, and to engage in physical activity to maintain a healthy body weight and BMI to prevent the onset of the chronic illness if possible.

LIVING WITH DIABETES

Some of the cardinal signs of diabetes, for both type 1 and 2 are: increased thirst, frequent urination, extreme thirst, unexplained weight loss, presences of ketones in the urine which cause sweet smelling urine, fatigue, irritability, blurred vision, slow-healing sores, frequent infections such as of the skin or vaginal infections (Center for Disease Control and Prevention, 2019). Untreated, these symptoms can worsen and also lead to a host of potential complications or further medical concerns. Some of these complications that can arise include but are not limited to kidney disease, foot and leg issues, eye disease such as retinopathy that can lead to blindness, heart attacks, nerve damage, and strokes (Diabetes Canada, 2021).

For individuals living with diabetes or are prediabetic, it is important to manage their symptoms in order to prevent further medical complications. For most individuals, with both type 1 and type 2 diabetes the most important measure is managing blood sugar and insulin levels. For most individuals this process involves utilizing a glucometer or a continuous glucose monitor that is capable of measuring blood glucose levels in a small sample of blood (Center for Disease Control and Prevention, 2019). Depending on the type of diabetes the individual has will determine how often they should check their blood sugar levels. Typical times that a diabetic would monitor their levels would be when they first wake up and before eating or drinking anything, before a meal, two hours after a meal, and again at bedtime. This may vary depending on the type or severity of diabetes in the patient, but for those more prone to low blood sugar levels a doctor may suggest monitoring on a more frequent basis.

Individuals with diabetes will often carry with them supplies or sugary beverages or candy to help treat hypoglycemia or low blood sugar. Hypoglycemic events can occur for diabetics as a result of missing a meal, taking too much insulin, exercising more than normal, as well as drinking alcohol (Center for Disease Control and Prevention, 2019). When an individual with diabetes is experiencing a hypoglycemic event, some signs can include shaking, sweating, irritability or confusion, dizziness and hunger although the presentation can be different for everyone. For diabetics, knowing the symptoms that they experience is important so they are able to prevent a hypoglycemic event from occurring if they feel one may present.

On the other hand, diabetics more often experience issues with high blood sugar or hyperglycemia. This can result from eating more than planned, as a result of a cold, or not giving themselves enough insulin (Center for Disease Control and Prevention, 2019). Some signs and symptoms of high blood sugar can include extreme thirst, fatigue, blurry vision, and the need to urinate on a frequent or more often basis.

A common difficulty for diabetics is being able to manage hypoglycemia during times of sickness as a result of not being able to eat food or drink fluids as they normally would. This can have a negative effect on blood glucose levels and even lead to diabetic ketoacidosis which is a more serious complication (Center for Disease Control and Prevention, 2019). For diabetics who are ill and suspect higher-than-normal blood glucose levels, it is recommended that they monitor the ketones in their urine – ketones are produced from the breakdown of fat when there isn't enough insulin present to allow blood sugar into your cells for energy consumption (Center for Disease Control and Prevention, 2019). Diabetic ketoacidosis (DKA) occurs when too many ketones are produced at a rapid rate and build up in your body and is characterized by common symptoms such as fast/deep breathing, dry skin and mouth, flushed face, sweet smelling breath, muscle stiffness or aches, stomach pain, and others. DKA is a serious consequence of diabetes and requires hospitalization to prevent more serious repercussions such as a coma or death.

Overall, it is crucial that individuals with diabetes pay close attention to their blood glucose levels through frequent and reliable monitoring. Diabetes and high blood glucose can lead to a whole host of difficulties and damage to other body functions and organ systems. Controlling and coping with diabetes can be difficult and distressing for individuals and as a result can cause diabetes distress. Individuals with diabetes are prone to feeling discouraged, worried, or even frustrated by the daily care that is required to treat their illness and may sometimes feel like they are controlled by their diabetes (Center for Disease Control and Prevention, 2019). As a result of the distress, some individuals may find themselves taking on unhealthy habits such as refraining from monitoring their blood sugar or ignoring their diabetes altogether.

In conclusion diabetes is a chronic illness that can affect anyone, but it can be prevented in many cases. Eating healthy and maintaining a healthy, balanced lifestyle is key to prevention or delay of diabetic onset for most individuals. To further understand diabetes and its effect on the human body, explore how the body and organ systems respond to the consequences of hyperglycemia. Though many individuals around the world are battling this chronic illness, with the help of doctors, counsellors, and endocrinologists it is a treatable condition that has been advanced by research and medicine.

6
SOCIETY'S VIEWPOINT

By Janani Rajendra

Statistics have shown that the prevalence of diabetes has been expeditiously increasing from 108 million people with diabetes in 1980 to 422 million people with diabetes in 2014 (World Health Organization, n.d.). Diabetes used to be classified as a disease that primarily affects adults, but recently there has been an increasing number of children being diagnosed with diabetes (World Health Organization, n.d.). Further, low income and middle income countries seem to be more affected than high income countries in regard to the increasing number of people being diagnosed with diabetes (World Health Organization, n.d.). The rise in diabetes cases is definitely a major concern in today's society as this disease is a major contributor to blindness, kidney failure, heart attacks, stroke and lower limb amputations (World Health Organization, n.d.). Moreover, an increase in premature mortality as a result of diabetes has been observed as well (World Health Organization, n.d.).

Most people with diabetes have type 2 diabetes (World Health Organization, n.d.). Type 2 diabetes occurs due to being overweight and leading a sedentary lifestyle with minimal physical activity (World Health Organization, n.d.). Type 2 diabetes usually occurs in adulthood, but recently more children have been diagnosed with type 2 diabetes (World Health Organization, n.d.).

Poor lifestyle and dietary choices play a significant role in developing type 2 diabetes (Tabish, 2007). Poverty can influence people to consume low cost-per-calorie foods and unhealthy foods ultimately predisposing them to type 2 diabetes (Tabish, 2007). Even in well developed countries, the poorest communities are seen to suffer greater rates of diabetes (Tabish, 2007). As such, economic advancement in both developed and underdeveloped countries can change diet and lifestyle, eventually leading to lower rates of type 2 diabetes (Tabish, 2007).

When one is first diagnosed with diabetes, they often feel devastated and displeased (EmmaHook, n.d.). As with being diagnosed with any disease, the initial feelings of hopelessness eventually fade away as the patient learns to cope with their feelings; however, sometimes these negative feelings take a very long time to get over (EmmaHook, n.d.). Patients with diabetes will have to modify their lifestyle and dietary intake in order to live a healthier life. As such, diabetes will affect the day to day living of the diabetic patient. These sudden changes in their life and their feelings of anxiety can sometimes lead to depression (EmmaHook, n.d.). Research has shown that individuals diagnosed with diabetes are twice as likely to develop depression compared to individuals without diabetes (EmmaHook, n.d.).

Regardless of what disease an individual has, they often face judgement from society. These judgemental comments and actions often stem from misunderstandings and false assumptions (Solomon, 2016). Many patients with diabetes are overweight and as a result, they can be criticized based on their body size and lifestyle choices. Diabetic patients often receive unsolicited advice from others on "how to fix [them]." (Solomon, 2016) Other biases such as race may play a role in how diabetic individuals are viewed in society, ultimately making it very difficult for these patients to get through their day happily (Solomon, 2016). Diabetic individuals can be treated poorly; for example, family members may put their diabetic family members on grueling diets and shame them for their body weight (Solomon, 2016). The conduct of others can severely affect patients causing them to feel ashamed about being diabetic (Solomon, 2016). This feeling can cause increased hopelessness beyond the agony they are already battling. Additionally, individuals diagnosed with diabetes at a very young age begin to diet in order to live a healthier life. Starting to diet at a young age can severely affect the mental well-being of the child, as they are not fully aware of what calories and maintaining blood glucose actually mean (Solomon, 2016). In extreme cases, diabetic children may overeat and then try to induce themselves to vomit, "I'd secretly drink sample bottles of perfume to try to make myself vomit." (Solomon, 2016)

Diabetes not only affects one's physical health, but it can also severely impact one's emotions (EmmaHook, n.d.). The stigma associated with diabetes causes newly diagnosed diabetic patients to hide their diagnosis from others due to the fear of being judged (Llamas, 2016). For example, a participant from a study said that she "felt the need to hide [her diagnosis] because she had a high ranking position at her company." (Llamas, 2016) Although some countries have laws that aim to prevent discrimination against diabetes as a disability, many diabetic patients still fear that stereotypes will lead to discrimination and restrict opportunities for them in the workplace (Llamas, 2016). Negative stigma associated with diabetes as well as the outward appearance of an individual (such as excess body weight) has been shown to affect opportunities and relationships (Llamas, 2016).

The media often portrays diabetes as associated with being overweight, lazy, and excessively consuming sweets (Llamas, 2016). The stigma caused by the media makes society believe that individuals with diabetes brought the disease upon themselves and that they would not have developed diabetes if they limited their sugar consumption and did exercise (Llamas, 2016). "Fat, obese, overweight, big fat pig, lazy, slothful, couch potato, over-eater and glutton" were words used to describe people with type 2 diabetes, according to a study conducted by Jessica L. Browne and colleagues which was published in 2013 in BMJ. The stigma arises from the fact that diabetes affects an individual's lifestyle and hence it is believed to be induced by laziness and lack of self-care (Llamas, 2016). Research has shown that more than half of Americans with type 1 and type 2 diabetes believe that others criticize them for being diabetic (Llamas, 2016). About 83% of parents with children diagnosed with type 1 diabetes believe that the public criticizes them for their child's type 1 diabetes (Llamas, 2016).

In today's society, the internet has become a popular source for digital media content (Llamas, 2016). Many people use social media networks such as Facebook, Instagram, TikTok, and YouTube to share moments of their life with the world and doing such has both positive and negative consequences. Mike Durbin who is a health blogger with type 2 diabetes and congestive heart failure shares his experience with negative online comments:

Some of the [online comments] were: 'If this guy would just get off his [couch] and do something, try exercising, try eating better, [he wouldn't have diabetes]' — most of the typical comments that you hear toward people with type 2," he said. "I've gotten to where I really don't take much of that to heart. It would really just eat you alive if you did. (Llamas, 2016)

Research has shown that people with type 2 diabetes felt blamed for their diagnosis when they saw content that emphasized their excess body weight and sedentary lifestyle as the reason for the disease (Llamas, 2016). On the other hand, the media tends not to portray individuals who are successful at managing their diabetes (Llamas, 2016). The great emphasis put on the negative aspect of diabetes makes it extremely difficult for those with type 2 diabetes not to feel guilty and ashamed about their diagnosis. "When I first got [diabetes] I wouldn't tell anybody. I didn't even tell my husband. I told nobody. I actually felt so ashamed to have diabetes. I felt completely ashamed of myself," says a 56 year old women with type 2 diabetes (Llamas, 2016).

Individuals with diabetes also felt judged and discouraged by health care professionals (Llamas, 2016). Patients often found that health care workers tended to focus on what the patient did "wrong", instead of being more encouraging and supportive (Llamas, 2016). Here is the experience of a 35 year old woman with type 2 diabetes:

The dietician was awful... she asked me if I exercise, and I said 'I do the gym twice a week and I have consistently since November.' 'That's not enough, you need to go five times a week.' This makes me really angry. (Llamas, 2016)

Dr. Sanjay Gupta says that doctors do have a tendency to believe that obesity is a self-created illness due to poor behavioral and lifestyle choices (Llamas, 2016). Further he claims that "some doctors will admit they are less sympathetic to their diabetic patients." (Llamas, 2016) "[My patient] was in the emergency room for a condition I considered completely preventable," says Dr. Peter Attia, a doctor that judged his type 2 diabetic patient due to his patient being overweight (Llamas, 2016).

Conjointly, patients with diabetes are often labeled as a "diabetic." (Llamas, 2016) Referring to individuals as diabetics is a "huge pet peeve" to many as it causes depersonalization and contributes to the stigma associated with diabetes (Llamas, 2016). Karen Kemmis, a diabetes educator at SUNY Upstate Medical University Syracuse, New York says:

We don't call someone who has cancer a canceric and shouldn't call someone with COPD a COPDer, But, somehow, it seems acceptable to many to call someone a diabetic. No, they have diabetes. We should use person-first language rather than label someone by a disease. (Llamas, 2016)

Although many family members and friends try to support individuals with diabetes, they sometimes end up being the "diabetes police"—an individual that constantly tells a diabetic person what they should and should not do (Llamas, 2016). Family members and friends were often described as "unhelpful, annoying or discouraging" according to individuals with type 2 diabetes (Llamas, 2016). Further, having diabetes makes it difficult for patients to be intimate with their spouse (Llamas, 2016). Around half of individuals with diabetes experience some type of sexual complication and this is an additional burden that individuals with diabetes face (Llamas, 2016). Diabetic individuals worry that others would not want to date someone with a lifelong disease that needs to be taken care of daily (Llamas, 2016). Even the act of telling their partner that they have diabetes can be quite intimidating to diabetic individuals (Llamas, 2016). Diabetes educator Janis Roszler provides suggestions on how to deal with diabetes:

Give 'diabetes' a name, and treat it like a totally separate entity," she said. "I know of a couple who named their husband's diabetes 'George.' When the husband started to feel his blood sugar drop during sexual activity, they blamed 'George' for making trouble, not the husband. (Llamas, 2016)

This method allows diabetic individuals in the relationship to feel less blame for their health condition.

Blood glucose levels can also be affected by stress; thus, it is important for diabetic patients to know how to manage their stress (Tabish, 2007). Due to the coronavirus pandemic, individuals from all over the world are suffering greatly from both physical and mental trauma. Simple everyday tasks, such as buying groceries at the grocery store, can become frightening during a pandemic due to the fear of being infected by the virus. Certain people such as those with diabetes are more vulnerable to being infected by the virus (EmmaHook, n.d.). This increased susceptibility can cause feelings of anxiousness and tension, leading to increasing levels of stress (EmmaHook, n.d.). Further, many diabetes appointments with health care professionals may be cancelled and medical supplies and drugs may be limited, adding to the hardship that diabetic patients face (EmmaHook, n.d.). Low inventory in stores causes many people to panic buy, resulting in little to no supplies left for those who really need them.

Diabetic patients are constantly trying to manage their diabetes by calculating the ratios of carbohydrates to fat to sugar, in order to determine the amount of insulin needed (Solomon, 2016). Even after performing and perfecting many calculations, the blood sugar levels can differ (Solomon, 2016). This can result in blood sugar levels falling too low (hypoglycemia) as a result of taking too much insulin (Solomon, 2016). To counteract this effect, one would try to eat more food such as an apple to raise blood sugar levels, but unfortunately blood glucose levels may shoot up again, well over the desired blood glucose levels, "Even the sugar in a serving of broccoli sends my sugars to uncomfortable highs." (Solomon, 2016) The constant stress over managing food consumption and blood glucose levels can make it extremely difficult for diabetic individuals to go to parties and restaurants (Solomon, 2016). This clearly demonstrates how difficult it can be for diabetic patients to monitor their blood glucose levels. Here are statements said by a diabetic patient:

The extremism with which I tackle diabetes management is directly related to the extremism I apply to food in general. A lifetime of dieting, a lifetime of being told my body is wrong, takes it toll, and I can't help conflating the messages that I am better off starved than fat. Maybe if I could let go of the shame, or more important, if the media, doctors, friends, family could stop shaming me, managing my diabetes wouldn't be this roulette wheel of self-torture. Maybe then, I could finally let go and heal. (Solomon, 2016)

Compassion and having someone to talk to can go a long way in helping a patient cope with diabetes. Expressing how one truly feels and the challenges they are facing can help relieve some stress (Centers for Disease Control and Prevention, 2019). In addition, family and friends can help take care and monitor one's diabetes (Centers for Disease Control and Prevention, 2019). For example, they can remind one to take medicine, help monitor blood glucose levels, help cook healthier meals, and provide emotional and moral support (Centers for Disease Control and Prevention, 2019). Talking to other individuals with diabetes can make one feel better as another person with diabetes would know what they are going through (Centers for Disease Control and Prevention, 2019). They can also provide advice and tips on how to better cope with diabetes (Centers for Disease Control and Prevention, 2019).

All in all, there is a lot of negative stigma associated with diabetes making the lives of those with diabetes quite difficult. It is important to educate oneself about diabetes as the rate of diabetes has been increasing rapidly. There are many programs available to help cope with diabetes and reaching out to family, friends, and health care providers can go a long way in helping one relieve their anxiety and stress caused by diabetes. Reaching out for help is an essential step that will make one's diabetes journey much easier than trying to get through it alone. Although being diagnosed with diabetes can be very intimidating and distressing at first, it is not the end of the world, and it does not represent a personal failing. There are many treatment options available as well as support programs available to help individuals manage their health condition.

7
SCIENTIFIC PERSPECTIVE

By Sudipta Samadder

Metabolism is the process by which the body converts food into energy. Notably, the digestive system converts carbohydrates, a type of macronutrient found in certain foods and drinks, into glucose that enters the bloodstream (NIH, 2021). Cells would then absorb the glucose with the aid of a hormone called insulin, which is produced by the beta cells in the pancreas. This would allow cells to convert the glucose into energy (NIH, 2021). If not enough insulin is produced by these beta cells, or if the body can't use insulin as well as it should, glucose will accumulate in the bloodstream leading to diabetes. Diabetes can cause severe health complications including heart disease, kidney failure, and amputations (NIH, 2021). As technologies have advanced over the last few decades, so have treatment options for patients with diabetes. The technologies involved in diabetes refers to the hardware, devices, and softwares that diabetic populations use to help manage their blood glucose levels, reduce complications of diabetes, alleviate the burden of living with diabetes, and improve the overall quality of life (American Diabetes Association, 2019). In the past, diabetes technology has been divided into two main categories: insulin administered by syringe, pen, or pump, and blood glucose monitoring as assessed by meter or continuous glucose monitor. More recently diabetes technology has expanded to include hybrid devices that both monitor glucose and deliver insulin, some automatically, as well as software that serves as a medical device, providing diabetes self-management support (American Diabetes Association, 2019).

INSULIN DELIVERY

Insulin Syringes and Pens
For diabetics who need insulin, a syringe or pen is usually the insulin delivery method of choice. The remaining population of diabetics prefer using insulin pumps or automated insulin delivery devices. Both insulin syringes and pens are capable of safely and effectively delivering insulin to attain glycemic targets (American Diabetes Association, 2019). Some factors that influence the decision of choosing between a syringe and a pen include patient preferences, self-management capabilities, cost, type of insulin and amount of dosage that can be delivered (American Diabetes Association, 2019).

An insulin pen has three parts: a needle, a barrel, and a plunger. The needle is short, thin and covered with a layer of silicone that helps it to easily pass through the skin without inflicting much pain. Additionally, a cap is used to cover the needle before use. The barrel component is the long chamber which stores the insulin. The plunger is used to slide up and down to either draw insulin into the barrel or push the insulin out through the needle (Healthwise, 2021). Most syringes have a volume of 1 mL, 0.5 mL, and 0.3 mL, allowing doses of up to 100 units, 50 units, and 30 units of U-100 insulin, respectively. In 2016, the FDA approved a U-500 specific insulin syringe (Kesavadev et al., 2020). The intended use of a syringe is normally one-time, but some can be reused by the same person in resource-limited settings if they ensure to clean and store the used syringe appropriately. Another consideration is the thickness (gauge) and length of the needle (Kesavadev et al., 2020). If the needle is thicker, the injection may be administered faster, while a thin needle may inflict less pain. A needle can be as long as 12.7 mm or as short as 4 mm, with shorter needles potentially lowering risks of intramuscular injection (Kesavadev et al., 2020). It is important for insulin injection therapy to be performed correctly so that it can provide the maximum benefit. To ensure correct and effective administration of insulin, a disposable patch-like device can be used (American Diabetes Association, 2019). It is important to note that needles that are reused may get dull and cause more pain during injections. This device provides a continuous, subcutaneous infusion of rapid-acting insulin, as well as 2-unit increments of bolus insulin by simply pressing a button (American Diabetes Association, 2019).

On the other hand, when insulin pens were introduced, they marked a new milestone in insulin delivery. If the patient has a disability or vision impairment, insulin pens or insulin injection aids may be recommended to facilitate the administration of accurate insulin doses (American Diabetes Association, 2019). The NovoPen was the first insulin pen that was introduced in 1985. This was followed by the introduction of NovoPen 2 in 1988 which had a dial-up setting to measure the required dose of insulin (Kesavadev et al., 2020). Insulin pens may be preferred over syringes due to the convenience of having the vial and syringe combined in a single device, which offers more convenience, simplicity, and accuracy (American Diabetes Association, 2019). These pens can also allow diabetics to inject insulin using a push button. The world's first disposable pen, known as the Novolet, was introduced by Novo in 1989 (Kesavadev et al., 2020). The insulin can be made available in disposable pens with prefilled cartridges or reusable insulin pens with replaceable cartridges. Another advantage of some reusable pens is the inclusion of a memory function, which can recall the amount of dosage that needs to be administered at specified times (Kesavadev et al., 2020). If patients would like a device that calculates insulin doses and allows access to downloadable data reports, "smart" pens would be the best option to serve that purpose. Pens can also differ by their dosage increments and minimum dose. Overall, they have better glycemic control than syringes and have wider acceptance (Kesavadev et al., 2020).

Insulin therapy varies between individuals living with Type 1 and Type 2 diabetes. While insulin is a requirement for individuals with type 1 diabetes, it is not always necessary for people with Type 2 diabetes (American Diabetes Association, 2021). Generally, people who have type 1 diabetes must inject two different types of insulin twice a day. On the other hand, people with Type 2 diabetes may require just one injection a day. Diabetes pills may stop working, so people with type 2 diabetes begin receiving two different types of insulin injections every day (American Diabetes Association, 2021).

Insulin Pumps

An insulin pump is a medical device that can be used as a substitute for more frequent injections of insulin using a syringe or pen (American Diabetes Association, 2019). Insulin pumps are becoming increasingly popular, although they are not any simpler to use than syringes or insulin pens. It is important to understand how to use pumps properly as patients can risk serious side effects from uncontrolled diabetes through improper use of pumps (Health Canada, 2012). In the majority of cases of type 1 diabetes, either daily insulin injections or insulin pumps should be used by adults, children, and adolescents. In the U.S., continuous subcutaneous insulin injections (CSII) or insulin pumps have been available for 40 years (American Diabetes Association, 2019). This type of device assists with controlling blood glucose levels throughout the day with rapid-acting insulin. A cannula is used to deliver insulin in most insulin pumps, but there are a few that connect directly to the skin, without a tube (American Diabetes Association, 2019). There are two types of insulin doses that are delivered to the body using a pump. One is basal insulin doses which are provided constantly for over 24 hours. It helps to keep the patient's blood glucose levels balanced and stable between meals (American Diabetes Association, 2019). Another type of dose is bolus insulin dose which is provided when the patient pushes a button on the pump. This feature can be used when the patient eats or when needs to correct high blood glucose levels. Another advantage of insulin pumps is that they can be worn in multiple places on the body (American Diabetes Association, 2019). For instance, it can be attached to the user's waistband, pocket, armband, or under garments (Health Canada, 2012). According to US studies, people with higher socioeconomic status are more likely to receive pump therapy. These factors include race/ethnicity, private health insurance, family income, and education (American Diabetes Association, 2019). Infusion set problems (dislodgement, occlusion) are often the cause of pump complications. Pump site infections may also occur.

Nasal Glucagon

One of the most severe complications associated with diabetes is hypoglycemia. Hypoglycemia is a condition in which blood glucose levels are lower than normal, and is often related to diabetes treatment (Pietrzak and Szadkowska, 2020). Hypoglycemia is still treated with intramuscular or subcutaneous injections of glucose or glucagon (a hormone that controls blood glucose levels). In 1983, intranasal glucagon drops were shown for the first time to elevate blood glucose levels in healthy volunteers (Pietrzak and Szadkowska, 2020). In following years, a new powder formulation of glucagon was developed, which was proven to be as effective as injectable glucagon. This powder formulation could be administered through the nose and passively absorbed through the nasal mucosa (Pietrzak and Szadkowska, 2020). In 2019, FDA had approved Baqsimi nasal powder, the first glucagon therapy for emergency treatment of severe hypoglycemia (FDA, 2019). Baqsimi increases blood sugar levels in the body by stimulating the liver to release stored glucose into the bloodstream. This is opposite to the effect of insulin, which reduces blood glucose levels (FDA, 2019). To use, one simply inserts the vial into the nostril, pushes a button and inhales a puff of the powder.

A Hybrid Closed Loop System

Hybrid closed loop insulin delivery systems (HCLs) are an emerging technology that combines an insulin pump with continuous glucose monitor (CGM) and a computer program (CADTH, 2021). The system uses information from the CGM to determine the amount of insulin to be administered throughout the day and helps keep a user's blood sugar within a predetermined range (CADTH, 2021). If properly used, these insulin delivery systems and monitoring can improve glucose control and reduce the risk of hypoglycemia (Neumiller, 2019). These devices represent one of the most advanced forms of insulin delivery, particularly for patients with type 1 diabetes. Since this system is only automated in the delivery of basal insulin, it is regarded as a "hybrid" closed-loop (Knebel and Neumiller, 2019). This means that users must still manually deliver bolus doses to cover meals or correct residual hyperglycemia. The 670G HCL system was initially approved by the FDA in 2017 for use in people ages 14 years or above (Knebel and Neumiller, 2019). In 2018, the system was expanded to include people with diabetes between the age of 7 and 13 years (Knebel and Neumiller, 2019). In a study of adolescents and adults older than 14, the use of the system resulted in improvements in blood glucose level, increased time spent within the glucose target range, and less overall hypoglycemia and hyperglycemia when compared to baseline (Knebel and Neumiller, 2019).

TRACKING GLUCOSE LEVELS

Blood Glucose Meters

Meters can be used to track blood glucose levels in the body. Two main types of meters are standard blood glucose meters and continuous glucose monitors (CGMs). The standard meters use a drop of blood to check a patient's blood glucose level at that moment (Lori et al., 2018). Whereas CGMs check blood glucose regularly, day or night. A basic understanding of how blood glucose meters work could be helpful when selecting one. The first step to using most blood glucose meters is inserting a test strip into the device (Lori et al., 2018). A patient would then use a needle to puncture a clean fingertip and obtain a drop of blood. The test strip would touch the blood droplet and the patient must wait for the blood glucose reading to appear on the screen of the glucose meter (Lori et al., 2018).

Even though finger pokes remain the gold standard for monitoring blood sugar, researchers are working on technologies that will make the process more convenient and less painful for the user. One example includes alternative site monitors (Mayo Clinic, 2019). These devices are designed to allow blood samples to be taken from areas in the body likely to be less painful than the finger. This may include the thighs, arms or even hand palm. It is worth noting that this process may not be as precise as obtaining fingertip samples especially when blood glucose level rises or falls too rapidly (Mayo Clinic, 2019). Continuous glucose monitors are also alternatives that involve placing a sensor under the skin to measure blood glucose level. The reading is transmitted to a small recording device and an alarm can be set to alert the user of very low or very high blood glucose levels (Mayo Clinic, 2019). CGMs are always turned on and recording glucose levels, whether the user is showering, working, exercising or sleeping.

Currently, there are four FDA-approved CGM systems on the market. The first official model device was Dexcom which continues to lead the market today. Dexcom's latest system, the Dexcom G6 CGM, offers zero-finger-stick technology, which eliminates the need for users to set baseline levels with finger stick tests (Brown, 2020). The FDA gave approval for the device to be used safely by children as young as two years old. The device allows users to customize alerts, and share device data with up to 10 followers, which can include the user's doctor, caregiver, and family members. Medtronic is also a leader in insulin pump technology and makes a CGM device called the Guardian (Brown, 2020). In 2018, the FDA approved the latest model known as the Guardian Connect, which is the company's first stand-alone CGM. The main advantage of Guardian Connect is protection, as it is the only CGM system that not only predicts where glucose levels are headed, but also alerts users 10 to 60 minutes before they experience a glucose excursion (Brown, 2020). This allows users to take the necessary action to avoid high and low episodes of their glucose imbalance. Currently, the Guardian Connect is available to use for users ages 17 to 75. The third company that started making CGMs were Abbot Diabetes Care in 2017 (Brown, 2020). The glucose monitor they designed was named the FreeStyle Libre Flash Glucose Monitor. What makes the system unique is that the sensor is quite small, can be scanned through clothing, and it is water-resistant which means users can put it on when swimming (Brown, 2020). The monitor had received approval for use for 18 years of age and older. Lastly, the last company which produces the world's first long-term implantable CGM is Senseonics (Brown, 2020). The device consists of a tiny sensor which is the size of a twig that is placed underneath the skin in the upper arm. In Europe, it is approved for 180 days of wear, whereas in the U.S it is approved for 90 days of wear, thus establishing it as the longest-lasting sensor (Brown, 2020).

Conclusion
Digital technologies for diabetes have been revolutionary in the treatment and monitoring of the disease within the last few decades. From blood glucose meters and continuous glucose monitoring (CGM), as well as introduction of different types of insulin delivery options, devices are easier to use and less invasive. These technologies have proven to help caregivers give better quality care at relatively low cost.

8

WHAT PUTS US AT RISK FOR DEVELOPING DIABETES?

By Megan Ng

INTRODUCTION

The World Health Organization (2021) states that there has been a rising trend in diabetes and there were an estimated 1.5 million deaths directly attributed to it in 2019. This makes it important to identify and be aware of the risk factors that can increase the chances of getting or advancing the severity of it. Diagnosis is a process which considers many factors contributing to the disease as a step towards determining the appropriate treatment. There are several types and stages of diabetes, which each have their own pathways of development (referred to as its pathogenesis). By understanding the pathogenesis of diabetes and the risk factors which contribute to it, the appropriate diagnostic procedures can be carried out to recommend treatments.

Diabetes mellitus (referred to as diabetes) is a condition where diseases create problems in the body's hormone insulin (refer to Chapter 5), and results in high blood sugar (hyperglycemia) and other health affecting symptoms (WebMD, 2021). The general pathogenesis of diabetes is when you eat food, the body breaks it down into sugars (glucose) in the bloodstream which induces the pancreas to release insulin (Dansinger, 2019).

The function of insulin is to remove glucose from blood to use as energy for the body, so when insulin function is interrupted there will be an excess of glucose in the blood (Dansinger, 2019). However, the way insulin is interrupted from this important job differs based on the type of diabetes.

TYPES OF DIABETES AND RISK FACTORS

Type 1 Diabetes (T1D) Risk Factors
Type 1 Diabetes (T1D) is a type of diabetes that has an earlier onset in life, has a lower prevalence of around 10% of all diabetes cases, and is also called insulin-dependent diabetes (Dansinger, 2019; Paschou et al., 2018). This means that the way insulin is disrupted in the body system is by the degradation of pancreatic ß cells which are in charge of producing insulin, and is thus characterized by hyperglycemia (high blood sugar) and a constant deficiency of insulin (Esposito et al., 2019). There are a variety of potential risk factors contributing to the breakdown of the ß cells.

Genetic factors are a risk factor for developing T1D and results in individuals with a genetic predisposition for T1D. For T1D, people who are relatives with a T1D individual have a higher chance of developing it at around 5 to 6% compared to the natural rate of occurrence in a population at around 0.4% (Paschou et al., 2018). This indicates that there may be genes which can be inherited to create this increased risk. MHC is a group of molecular locations on genes (loci) that are responsible for immune cell recognition of foreign substances (antigens) by cells called T lymphocytes in the process of creating antibodies, a protection substance produced by the body to destroy foreign substances that may harm the body. In T1D individuals, the relevant genes to MHC show a different version than a normal person (an allele), which affects the ability for the T lymphocytes to bind to the appropriate ß cells which create these antibodies. Another gene component connected to this is insulin-VNTR, which are sections of the insulation gene promoter which have changed to promote deletion of T lymphocytes instead of producing them (Paschou et al., 2018). To simplify what this means, T1D is an autoimmune disease where the immune system mistakes its own body as a target for attacking substances like antigens, and with the interference of MHCs, insulin-VTR, and many other gene variations, T1D patients have an immune system that attacks their own pancreatic ß cells or other insulin producing functions (Paschou et al., 2018). This nature of this attack is suggested to be apoptosis, which is programmed cell death, induced by the T lymphocytes when they need to bind to ß cells. This highlights the importance of regular screening for relatives of T1D patients to identify people at risk for developing T1D.

Epigenetic factors, which are heritable changes in the expression of genes with no change in the underlying DNA sequence, are also a factor influencing the risk of T1D (Esposito et al., 2019). These are important risk factors in the process of diagnosis due to the invention of epidrugs, which is medicine that can reverse the epigenetic variations to prevent T1D (Esposito et al., 2019). DNA hypermethylation is one epigenetic variation that can be identified in T1D patients, where their genetic sequence has an excess of methyl groups bound to it (Esposito et al., 2019). Hypermethylation is associated with gene silencing, which can be considered to be a process that blocks the genetic sequence from doing what it is coded to do (Esposito et al., 2019). This means that inappropriate methylation at certain sites will induce the development of insulin autoantibodies (self-attacking antibodies) rather than regular antibodies, which do not attack insulin producing functions (Esposito et al., 2019). Another epigenetic modification is miRNAs, which are adjustments to regulatory bodies for gene expression. MiRNAs have several methods of blocking insulin production, including blocking the transmission of ß cells to the appropriate places in the body and inducing an inflammatory damaging response to ß cells (Esposito et al., 2019). However, epidrugs are still under development in clinical trials, so it is also important to identify other environmental factors which T1D patients and potential individuals can adjust within their regular life (Esposito et al., 2019).

T1D development is also influenced by environmental factors like viruses, diet, and gut microbiota. Gut microbiota are an important part of the immune system, and changing its composition can create immune complications like T1D (Dedrick et al., 2020). In a DIABIMMUNE study, it was identified that infants progressing towards T1D had less diversity in the gut microbiome which can suggest less protection towards invasion of the body(Dedrick et al., 2020). Less bacteria gene content was also identified, and the remaining genes had more genes focused on transporting sugar in the blood rather than those for breaking down the sugar (Dedrick et al., 2020). Recent studies support that T1D pathogenesis involves the gut microbiota in terms of the reduction of the ratio between two species of bacteria, firmicutes and bacteroidetes (Paschou et al., 2018). This lowered immunity through the gut microbiota is affected by diet and use of antibiotics as well.

Breastfeeding is important to the development of an infant's gut microbiome since breast milk contains maternal microbiota that adjust the infant's microbiota to give it protection (Dedrick et al., 2020). Several studies indicate short breastfeeding durations before switching to other milks can become a risk factor later in life, but these studies have now been found to not have a clear association for other milks increasing negative effects or breast milk increasing positive effects (Esposito et al., 2019). This is still an area to be investigated in the study of diabetes, but is a potential risk factor to be considered in the development of a protective gut microbiome for T1D prone individuals. Another food which has been linked to potential protection against T1D is vitamin D. Although there are no conclusive studies, there has been evidence of reduced risk when administering doses of vitamin D regularly in young children at a particular stage of their development (Esposito et al., 2019). These dietary recommendations are important to consider in children with known genetic predisposition to T1D for prevention.

Along with dietary consideration, antibiotics also affect the gut microbiome. Antibiotics are medicine that target and control bacterial infections to the body (Dedrick et al., 2020). As a result, exposure to antibiotics has been found to reduce the number of bacterial species (diversity) in the microbiome (Dedrick et al., 2020). However, rather than finding more T1D in those who took courses of antibiotics early in life, studies found that mothers taking antibiotics before or during pregnancy could reduce the diversity of gut microbiota of their children and expose them to a higher risk of T1D (Dedrick et al., 2020). It was also found that several courses (5 at least) of an antibiotic are needed to create a significant increase in T1D risk (Dedrick et al., 2020). This emphasizes a risk factor to consider, especially in T1D individuals, when they choose to have children.

Type 2 Diabetes (T2D) Risk Factors

Type 2 Diabetes (T2D) is the other form of diabetes that is prevalent for about 90% of all diabetes cases, and is also called an adult onset, non-insulin-dependent version of diabetes (Dansinger, 2019). Unlike T1D, T2D presents hyperglycemia and insulin deficiency mostly due to a resistance to insulin rather than lower production of insulin (Dendup et al., 2018). Some T2D patients can still produce insulin, and their insulin resistance is developed through genetic, age, and lifestyle factors (Dansinger, 2019; Dendup et al., 2018). (71)

Prediabetes is used to describe people who are at a higher risk for developing T2D, and is very prevalent with more than 50% of American adults having it (Centers for Disease Control and Prevention, 2020; Rooney et al., 2021). Although people with prediabetes do not show diabetes complications, it is still important to know what factors put someone at this risk for the sake of prevention of progression to diabetes (Centers for Disease Control and Prevention, 2020). Prediabetes individuals are also at risk for heart disease, strokes, and other cardiovascular conditions (The Canadian Diabetes Association, 2018). The Centers for Disease Control and Prevention (2020) listed the following risk factors: those with direct family with type 2 diabetes, previous history of gestational diabetes, being overweight, being over the age of 45, and not being physically active at least 3 times a week. The pathogenesis is similar to T2D since it is considered a phase T2D patients pass through before developing diabetes (Bhansali & Dutta, 2005). It is characterized with high blood glucose levels after a meal, but lower early insulin response and delayed secondary secretion of insulin (Bhansali & Dutta, 2005). The diagnosis methods for prediabetes are blood tests during diabetes screenings, and the difference between diagnosing prediabetes and T2D is the level of blood sugar measured since prediabetes will have slightly lower levels (Mayo Clinic Staff, 2020). Those with prediabetes may be able to control their condition through lifestyle changes since this is still T2D and not T1D where the pathogenesis is not as affected by them. The prediabetes levels identified by Mayo Clinic (2020) are between 5.7% and 6.4% on a glycated hemoglobin test or 100 to 125 mg/dL on a fasting blood sugar test. However, these values vary based on ethnic, age, and gender factors as well, so it is necessary in specific populations to determine how to diagnose prediabetes (Abdul-Ghani & DeFronzo, 2009).

The pathogenesis of T2D begins with an increased production of insulin to bring the blood to a regular glucose level (Dendup et al., 2018). However, this level of insulin from the ß cells will not overcome the insensitivity resistance such that the liver continues to produce glucose, and the metabolism of macromolecules such as proteins and carbohydrates will thus be affected (Dendup et al., 2018). This results in the continual progressive damage to ß cell causing dysfunction (Dendup et al., 2018). This indicates that reducing energy-dense foods and increasing physical activity to reduce initial glucose levels causing the development of T2D are essential lifestyle changes to improve T2D patient conditions (Dendup et al., 2018).

Diet habits are again a risk factor for T2D, but unlike T1D, there is a focus on decreasing the high energy density foods that are regularly intaken and changing the types of meal styles rather than early interventions at the infant (Kolb & Martin, 2017). By reducing high energy density foods in the diet, theoretically, this will lower the blood sugar levels and be more protective against T2D, but nuts are a high energy density food, but eating a handful a day provides some protection against T2D (Kolb & Martin, 2017). There are several food patterns that have been identified such as eating more plants and fermented dairy products that are also beneficial (Kolb & Martin, 2017). Rather than providing set carbohydrate, fat and protein guidelines, it has appeared to be more effective to set food types and suggested cuisines that have shown protective abilities against T2D (Kolb & Martin, 2017). An interesting trend is several studies have found alcohol consumption at a moderate dosage (1 or 2 drinks) to show protection from T2D development in certain populations (Kolb & Martin, 2017). A study also found lower T2D prevalence in areas closer to healthy restaurants and grocery stores (Dendup et al., 2018). However, many studies in this area have been conducted and there are a variety of opinions and options which have been suggested. It would be difficult to summarize a consensus between these, but they all share that there are certain dietary habits which may improve T2D protection.

Physical activity has been a main recommendation for T2D patients since increased mass to the point of obesity and a sedentary life has been linked to this condition. Dendup et al. (2018) explains that accumulation of fat from physical inactivity can lead to ß cell dysfunction. The benefits of exercise for T2D is its ability to reincrease the individual's sensitivity to insulin with the appropriate weight loss (Kolb & Martin, 2017). Studies have indicated that combining physical exercise with a healthy diet can decrease the risk for T2D, and physical activity in itself has an effect of up to 30% risk reduction (Dendup et al., 2018; Kolb & Martin, 2017). On the opposite end, sedentary life (a low activity, seated lifestyle) has been strongly associated with developing obesity and T2D even if the individual does conduct in physical activity otherwise (Kolb & Martin, 2017). There are recommendations of even just standing or walking more as it expends more energy than simply sitting (Kolb & Martin, 2017). To connect this back to the pathology of T2D, lifestyle changes such as diet and exercise may be connected to the epigenetic DNA methylation patterns as well to influence genes connected to T2D similar to how they influence insulin in T1D (Zhou et al., 2018). Therefore, DNA methylation patterns are now also a potential way to diagnose T2D and track its progress (Zhou et al., 2018).

There are also the risk factor of age in developing T2D with the gradual breakdown of functions in the body such as decreased production of ß cells, as well as increased sensitivity to apoptosis (Halim & Halim, 2019). The peak age for T2D prevalence in the older population is ages 60-74, and is often predicted to arise with many other metabolic complications like nonalcoholic fatty liver and cardiovascular diseases with the aging of the body (Halim & Halim, 2019). The Canadian Diabetes Association (2021) suggests individuals above 40 to get tested every 3 years, and more frequently if they have more confounding risk factors.

Gestational Diabetes

Gestational diabetes is a unique form of diabetes that arises during pregnancy, limiting it to women only, and is identified by glucose intolerance (hyperglycemia) as well but without need for other typical symptoms (Bruns et al., 2020). This is an important condition to identify since it can create risks to the patient's pregnancy with increased chance of needing Caesarean delivery and increased birth weight of the baby (Bruns et al., 2020). Gestational diabetes shares many of the main risk factors listed above such as lifestyle practices in food and exercise. An umbrella review of many studies on gestational diabetes found that the main shared factor that presented strong evidence was the body mass index (BMI) of the mother before pregnancy (Giannakou et al., 2019). Several studies identified that both pre-pregnancy weight as well gestational (during pregnancy) weight gain needs to be controlled at an appropriate level, not underweight or overweight, to reduce overall risk to both the pregnancy and the mothers gestational diabetes (Najafi et al., 2019; Zhao et al., 2017). It is important for pregnant women to be aware of the risk of developing gestational diabetes as well as the care that needs to take place if they do have it.

Although BMI is not the only factor, it is difficult to strongly identify many risks due to the complicating factors of the state of pregnancy.

CONCLUSION

In conclusion, depending on the type of diabetes there are many different and overlapping risk factors that can be identified which contribute to the development and progression of diabetes. The list of factors above is far from comprehensive, but highlights several heavily researched and focused on risks. What can be taken away from knowing these risks is recognizing these risks need to be checked to prevent more severe consequences, and some of these risks can actually be reduced through the affected individuals efforts.

9

OPPOSING VIEWPOINTS, CONTROVERSY, PUBLIC OPINION

By Maria Gonzalez

Although a plethora of peer-reviewed and scholarly resources exist about practically any topic one might feel curious about, misinformation is also abundant and is therefore susceptible to being spread. In order to combat this misinformation, it is important for one to do their own research, verify facts using reputable sources, and develop an opinion that represents their values and morals. It is particularly important for people to consume accurate, thoroughly researched information about medicine, medical disorders, and other topics in the fields of health and wellness which have a direct impact on their livelihoods. Diabetes is one of such topics. The disorder has been a source of controversy and dispute because of the confusion and misconceptions surrounding its risk factors, treatments, and who it can affect. Diabetes is a disorder unbound to age, sex, or race, and because of its rising prevalence in the world's population, it is vital to dispel the myths and misinformation surrounding it. Additionally, public opinion on diabetes can be swayed by the attitudes of health care professionals when treating patients with diabetes. Correcting misinformation about diabetes can reassure people of the treatment options available to them, increase trust in their health care providers, and protect them with knowledge of the legitimate risk factors of diabetes.

The idea that sugary foods and their overconsumption are the cause of diabetes is not an uncommon one. Because it is so widely believed, it is one of the largest misconceptions about diabetes that exists. As described in earlier chapters, diabetes has three subtypes—type 1, type 2, and gestational (Mayo Clinic, 2020). Type 2 diabetes is the subtype people are ordinarily thinking of when they misidentify sugary foods as the culprit behind the development of diabetes, because type 1 is typically diagnosed in childhood or adolescence, and type 2 is the most common type overall (Mayo Clinic, 2020). The likely reason for this misconception is the elevated blood sugar levels associated with diabetes (Mayo Clinic, 2020). It would therefore not seem unreasonable to suggest that eating a diet laden with sugar would in turn increase blood sugar levels, but this is not necessarily the case. Although overconsumption of sugar is a partial contributor to the development of type 2 diabetes, it is neither the direct nor sole cause. The onset of type 2 diabetes is often attributed to poor dietary habits, which can include the overconsumption of sugar, but concern should instead focus on excess weight; being overweight or obese is a legitimate risk factor for developing diabetes, and this may or may not result from consuming a diet high in sugar (Mayo Clinic, 2020). It should be stated that although most people with type 2 diabetes are overweight—specifically 89% of Americans, according to a 2020 statistics report from the CDC—not all are (Centers for Disease Control and Prevention, 2020). Incorrectly attributing diabetes to sugar consumption alone is harmful because it ignores the overarching issue of obesity and excess weight and other risk factors in contributing to diabetes. In saying so, the findings of studies attempting to link diabetes and sugar consumption—particularly sugary beverage consumption—are not always in alignment.

In one study conducted on a population of roughly 70,000 Mexican women over a period of 7 years, the researchers aimed to establish a link between sugary beverages—specifically soda—and the development of diabetes. They also questioned whether the potential link between sugary beverages and diabetes could be impacted by genetic dispositions to diabetes or early life factors (Stern et al., 2019, p. 795). Diet, and consequently sugary beverage intake, as well as diabetes diagnoses and treatment, was assessed using a questionnaire (Stern et al., 2019, p. 796). Early life factors and genetic predispositions to diabetes were also assessed using a baseline questionnaire with questions about family history, socioeconomic status, lifestyle, birth weight, and physical activity levels, among others (Stern et al., 2019, pp. 796-797). Stern et al. (2019) found that, after adjusting for confounding variables, the consumption of sugary beverages was associated with a higher incidence of diabetes (p. 800). It was also found that the relationship between sugary beverage consumption and diabetes was stronger in women that were overweight in childhood or young adulthood (Stern et al., 2019, p. 800). One possible explanation for this, according to Stern et al. (2019), is that women that were overweight in childhood had poor eating habits for a longer time period, which may have increased their susceptibility to developing diabetes later in life (p. 800). Stern et al. (2019) suggest that sugary beverages cause spikes in blood glucose and insulin levels, leading to a high glycemic load, which is linked to a higher risk of diabetes (p. 800). Furthermore, evidence indicates that consumption of sugary beverages made with fructose increases visceral fat growth, which raises insulin resistance, leading to a higher risk of developing diabetes (Stern et al., 2019, p. 800). Fructose may pose a problem—countries that regularly use high fructose corn syrup in their food have a 20% higher rate of diabetes than countries without high fructose corn syrup in their food (Stern et al., 2019, p. 800). Regarding Stern et al.'s (2019) objective of determining whether family history of diabetes would impact the link between sugary beverage consumption and diabetes prevalence, the researchers found that there was little difference in the magnitude of that link among those with family history of diabetes and those without (p. 800). In comparing the participants that more frequently drank sugary beverages with those that least frequently drank sugary beverages, it was found that the frequent consumers of sugary beverages were more likely to be obese, were younger, and "less likely to be in the highest socioeconomic status or physical activity categories" (Stern et al., 2019, p. 797). To sum up the findings of the study—sugary beverage consumption was in fact associated with an increase in prevalence of diabetes, but was also associated with excess weight.

Conversely, in a study conducted in Japan on a population of 2,037 middle-aged men, it was found that sugary beverages—sugar sweetened beverages (SSB)—were not associated with an increase in diabetes incidence (Sakurai et al., 2014, p. 256). Unlike the Stern et al. study discussed earlier, the Sakurai et al. (2014) study examined the link between diet soda and diabetes in addition to the already established link between SSB and diabetes, this time among a population of lean men to account for the body mass index (BMI) variable (p. 262). The study was conducted over a period of 7 years, where diabetes development was assessed at yearly physicals, and SSB and diet soda consumption information was collected via a self-administered questionnaire (Sakurai et al., 2014, p. 252), similar to the methodology of the Stern et al. study. Although the study found SSB consumption was not associated with diabetes in lean men, it did find that diet soda consumption was significantly associated with diabetes prevalence in lean men (Sakurai et al., 2014, p. 256). Furthermore, the results showed that, "participants who consumed at least one diet soda serving per week had a 70% greater risk of diabetes compared to those who did not consume diet soda" (Sakurai et al., 2014, p. 256). The reason why SSB consumption was not associated with diabetes is that when the results were adjusted to account for variables such as genetic predisposition to diabetes, weight gain or loss, and the use of diuretics, there was not a significant enough link (Sakurai et al., 2014, p. 256). There were similar confounding factors for the association between diet soda consumption and diabetes, such as dietary intervention for medical concerns like hypertension or chronic diseases—however, after adjusting for the confounding variables, the results still suggested statistical significance (Sakurai et al., 2014, p. 257). Another explanation the researchers propose for the lack of a significant link between SSB consumption and diabetes in their study is that their participants were generally younger and weighed less than the participants in previous studies and meta-analyses (Sakurai et al., 2014, p. 257). As mentioned previously, BMI—or excess weight—is a risk factor for diabetes, so it skews results and reduces the significance of any link between SSB consumption and diabetes found in studies attempting to establish one. Additionally, a link between excess weight and diet soda consumption is possible; the artificial sweeteners used in diet soda can increase cravings for sweet foods or foods high in energy, and people may overestimate how many calories they saved by choosing a diet soda over a SSB and then over consume other foods (Sakurai et al., 2014, p. 257). An important takeaway from the Sakurai et al. (2014) study is that diet soda consumption is not as effective at preventing diabetes as people may believe (p. 257).

The risk factors or causes of diabetes are not the only topic the public is misinformed about. Correcting false information spread about diabetes can improve how people view treatments used to manage the disorder, and also has the potential to encourage people recently diagnosed with diabetes to accept the best treatment option for themselves. There are many myths that surround the use of insulin therapy—a treatment option for diabetes described in chapter 7—and dispelling these myths is vital to making people more comfortable with and informed about diabetes in general, particularly if faced with a diagnosis in the future. People might be reluctant to accept or understand insulin therapy because of the preconceived notion that it can cause serious health complications. A 1997 study conducted on a population of 44 Mexican Americans diagnosed with type 2 diabetes aimed to uncover their perceptions about insulin (Reid, 2007, p. 183). The study found that 75% of the comments the participants made about insulin were negative in nature, with the most frequently cited reason being the possibility of health complications after starting insulin treatments (Reid, 2007, p. 183). The complications the participants listed included weight gain, energy loss, and kidney failure, among others, with blindness being the most commonly mentioned (Reid, 2007, p. 183). Concerns about insulin therapy can also arise from health care professionals—some of their concerns include: fearing the patient will be non-compliant, fearing the patient will gain weight, fearing they will drive the patient away to another medical care provider, and fearing the patient's anger (Reid, 2007, p. 183). While these fears are reasonable, it is important for health care providers to feel confident in their ability to treat diabetes and to trust that their patients will be receptive to treatments, regardless of what type. It is understandable to prescribe a different treatment when one causes undesirable or debilitating side effects, but to avoid prescribing a certain treatment because of fears of non-compliance or losing patients to other providers is an injustice to the patient. Building trust and assuaging fears by listening to the patient's concerns is the key to creating the optimal treatment plan for the patient (Reid, 2007, p. 187). One reason people believe insulin therapy will lead to serious health problems in the future is that they have heard horror stories from friends and family about bad experiences with insulin in the past (Reid, 2007, p. 184). Reid (2007) argues that the stories relayed by friends and family may sometimes not contain the entire truth—the health complications they faced were likely already developing prior to starting insulin therapy, and were not directly caused by the insulin itself (p. 185). Moreover, insulin therapy is prescribed much earlier now than in the past decades, in order to prevent those same exact health complications from occurring (Reid, 2007, p. 185). Reid (2007) suggests asking the patient about whether they have any family members with diabetes in order to help them shed any misconceptions formed by the stories they have heard in the past (p. 187).

Some other myths about insulin therapy are that it causes weight gain, or that it is an expensive treatment option. Weight gain is a valid concern for health care providers and their patients alike, as many people with type 2 diabetes already struggle with excess weight (Reid, 2007, p. 185). Weight gain is also not an impossible outcome, because "insulin does promote the movement of carbohydrates into metabolically active tissues such as muscle and adipose cells" (Reid, 2007, p. 185). Reid (2007) suggests that weight gain can be attributed to factors other than insulin therapy, such as patients not exercising regularly or preemptively overeating in order to combat hypoglycemia (p. 185). Furthermore, other treatments for diabetes—like oral agents—can also induce weight gain, so weight gain is not an exclusive side effect of insulin therapy (Reid, 2007, p. 185). In regards to the expensiveness of insulin therapy, Reid (2007) explains that increased testing to monitor blood glucose levels after starting insulin therapy would raise the cost of testing, but in general the prices of insulin are comparable to other treatments (p. 186). One analysis of patient records from 1177 people with type 2 diabetes found that 9 months after starting insulin therapy, health care costs had increased significantly in comparison to health care costs prior to starting insulin therapy (Reid, 2007, p. 186). However, the costs were shown to increase the most over the first 2 months, prior to dropping significantly—by 57%—for the remainder of the course of treatment (Reid, 2007, p. 186). In another study with a population of 188 people diagnosed with type 2 diabetes, the costs of two different treatment options were compared: three oral agents or two oral agents in addition to insulin (Reid, 2007, p. 186). The health care costs evaluated included medications, needles and training, and liver function tests; other costs, like dietary counseling or supplies for monitoring blood glucose levels, were assumed to be equal when comparing the two groups (Reid, 2007, p. 186). The researchers found that both treatment options (three oral agents or two oral agents with insulin) were equally as effective at treating diabetes, but the insulin treatment group was only a third of the cost of the other group (Reid, 2007, p. 186). As Reid's arguments suggest, insulin therapy is an effective and well monitored option for treating diabetes, and is not less worthwhile than other treatments.

To conclude, the spread of misinformation through anecdotes and poorly researched articles is dangerous, particularly in the fields of medicine, and health and wellness. Misinformation is especially dangerous when it is perceived as fact and begins to impact the health and safety of actual people. One such instance is the vast number of misconceptions surrounding diabetes. Although there are many scholarly resources available online or at libraries with accurate information about diabetes, myths like the consumption of sugar causing diabetes or insulin therapy causing serious health complications have become commonplace. Through studies on diverse populations, researchers and health care professionals have been able to dispel myths and correct misinformation about diabetes, its risk factors, and its treatments. Correcting misconceptions about medical conditions in general—not just diabetes—can reduce fear and encourage people to seek and accept medical care when needed.

10
FUTURE DIRECTIONS

By Tara Y.T. Chen

INTRODUCTION

It is estimated that by 2045, 10% of the global population will have diabetes (Ellahham, 2020). Type 1 diabetes (T1D) is caused by autoimmune destruction of insulin-producing beta cells in the islets of Langerhans in the pancreas (Pathak et al., 2019). Most type 2 diabetes (T2D) patients can produce insulin, but the cells in their body do not respond effectively to insulin, which results in high blood glucose levels (Lingvay et al., 2021). In the past four decades, the number of T1D and T2D cases have increased worldwide, resulting in increased research of more effective therapeutic techniques for diabetes. For T1D, research into orally-delivered insulin systems, advancements in artificial pancreases and islet transplantation are techniques with potential for future use in clinical settings (Pathak et al., 2019). The development of weekly insulin injections and insulin inhibitory receptor targeting are also emerging areas of research that may support T2D patients (Ansarullah et al., 2021). The use of artificial intelligence may produce more accurate diagnoses and patient-specific treatment plans that are predicted to improve the efficiency of healthcare systems (Ellahham, 2020).

ORALLY-DELIVERED INSULIN FOR T1D

Insulin replacement therapy is a common method for managing the amount of insulin present in a T1D patient. It involves the injection of exogenous insulin, which is insulin that has been produced outside of the body (Pathak et al., 2019). Recent research has led to the development of orally delivered insulin through an ingestible millimetre-scale device that can self-orient within the gastro-intestinal tissue. This is a non-invasive treatment that eliminates the need for needle injections of insulin. Results from diabetic rodent studies that utilized the orally delivered insulin device demonstrated stable insulin levels and normal blood glucose levels. It is proposed that future application of this orally delivered insulin device in humans may provide a beneficial alternative for insulin replacement therapy that avoids the use of frequent injections (Pathak et al., 2019).

ARTIFICIAL PANCREAS FOR T1D

Maintaining normal blood glucose levels in T1D patients without causing recurring periods of low blood glucose levels is a difficult task. Although insulin replacement therapy is commonly used, exogenous insulin is not able to completely replicate the biological behaviour of insulin that is normally produced in the body. Additionally, insulin replacement therapy leads to a lifelong dependency on daily insulin injections, dangerous periods of low blood glucose levels, insulin resistance and mental illnesses (Pathak et al., 2019).

An artificial pancreas is an emerging treatment method that incorporates continuous glucose monitoring and provides an appropriate insulin release rate to the body. This allows for better control over blood glucose levels, a reduction in insulin spikes and fewer periods of low blood glucose levels (Pathak et al., 2019). Moreover, an alternative artificial pancreas system called a dual-hormone system allows for even more precise management of T1D. This dual-hormone system administers both insulin and glucagon to the patient, which allows for tighter glucose regulation and more accurate replications of the physiological behaviour of a pancreas compared to the single-hormone artificial pancreas system (Pathak et al., 2019).

Regardless of the use of a single-hormone or a dual-hormone system, artificial pancreases have more efficient glucose control compared to the commonly used sensor-augmented insulin pumps and injections. Since the development of artificial pancreas systems is quite recent and studies exploring their use is limited, future research may utilize larger sample sizes and conduct longer follow-up durations to produce more reliable and accurate results. The high cost and potential need for sensor replacements in an artificial pancreas may also be investigated in the future (Pathak et al., 2019).

ISLET TRANSPLANTATION FOR T1D

Islet transplantation is a method of providing insulin to T1D patients without the need for exogenous insulin injections. In the past several decades, there have been islet transplants that resulted in effective regulation of blood glucose levels and management of T1D symptoms. However, the effectiveness of islet transplantations is hindered by possible islet cell death, issues with the patient's immune responses and a decrease in blood supply to the tissues. There are also limitations in the amount of donor islets available for transplant surgeries (Pathak et al., 2019).

Fortunately, there are emerging techniques that aim to minimize the dependency on islet cell donors, protect the newly transplanted islets from being attacked by the patient's immune response and maximize the amount of surviving islet cells being transplanted. These techniques utilize an encapsulation approach that consists of pancreatic progenitor cells enclosed in a delivery device (Pathak et al., 2019). The progenitor cells originating from the patient's stem cells have fewer negative immune responses compared to transplantations of donor islet cells. The progenitor cells would further differentiate into islet cells when placed into the patient's body. However, due to the device that surrounds the transplanted pancreatic progenitor cells, there is a physical barrier that minimizes the amount of oxygen that reaches the beta cells in the islets, which results in a beta cell death rate between 5% to 47% (Pathak et al., 2019).

Recently, the issue of insufficient oxygen delivery to the site of transplanted islet cells has been investigated. A device called βAir incorporates a refillable oxygen tank into the original encapsulation device design. The results from a study of four patients who received βAir device transplants consisting of 155 000 to 180 000 islet equivalents revealed complete beta cell survival for up to six months after surgery (Pathak et al., 2019). Unlike the original device design, the addition of a refillable oxygen tank allowed for sufficient oxygen delivery to the beta cells in the islets, which minimized the amount of beta cell death. However, the βAir transplantation study did not control blood glucose levels in the patient and had minimal long-term therapeutic impact (Pathak et al., 2019).

Although there are limitations in the βAir device, it provides insight into the development of islet encapsulation devices that prevent immune rejection of beta cells. For instance, recent advancements have utilized hydrogel to encapsulate rat islet cells, which prevented tissue damage and inflammation. It also maintained proper blood glucose levels in diabetic mice for up to four weeks. It is projected that there will be further investigation into the development of long-term therapeutic islet transplantation devices in the near future (Pathak et al., 2019).

WEEKLY INSULIN INJECTIONS FOR T2D

Similar to T1D patients, some T2D patients require daily insulin injections. However, millions of patients have a fear of injections and consider them an inconvenience. In addition to the development of orally-delivered insulin techniques, clinicians believe that providing an option for weekly insulin injections will decrease patients' reluctance in using insulin therapy. It may also be beneficial for patients with memory problems and decrease the need for daily insulin injection assistance for elderly T2D patients (Bajaj et al., 2021).

Insulin icodec is a novel insulin analog that can be administered weekly to diabetic patients. The effect of insulin icodec was analyzed in a 23-week study involving 205 T2D patients from the United States, Germany, Poland, Spain, Hungary, Croatia and Slovakia. An initial screening of patients occurred in the first two weeks, followed by weekly insulin icodec injections for 16 weeks. Afterwards, a five-week follow-up occurred to evaluate blood glucose levels. Different methods were also considered to adjust the amount of insulin dosage to establish a more stable and normal blood glucose level (Lingvay et al., 2021).

In a second study conducted by the same researchers, 154 T2D patients from the United States, Germany, Italy, the Czech Republic and Canada were administered weekly insulin icodec injections that followed the same study timeline as the first study. In this second study, there was a focus on determining the best method for patients to transition from daily to weekly insulin injections (Bajaj et al., 2021).

From the two studies, it was determined that weekly insulin injections with the novel insulin icodec analog was an effective method for maintaining proper blood glucose levels in T2D patients (Lingvay et al., 2021). In the second study, patients who had a higher initial insulin dosage reached their ideal blood glucose target faster compared to patients who had a lower initial insulin dosage. As a result, T2D patients who may switch from daily to weekly insulin injections should begin by administering weekly injections with a higher insulin dosage (Bajaj et al., 2021). Although the two studies focused on T2D patients, future studies may evaluate the effect of weekly insulin icodec injections in T1D patients. Since the results of the two studies were published recently in April of 2021, the long-term effects of using weekly insulin injections should be further investigated to determine the possibility of replacing daily insulin injections and standardizing weekly injections as a primary treatment method (Bajaj et al., 2021).

INSULIN INHIBITORY RECEPTOR TARGETING FOR T2D

The recent discovery of an insulin inhibitory receptor has the potential for enabling further advancements in diabetes research and treatment. Preventing the function of insulin inhibitory receptors can increase the sensitivity of insulin signalling pathways in beta cells in the pancreas. Consequently, beta cells are protected and potentially regenerated, which decreases diabetes symptoms in patients (Ansarullah et al., 2021).

Through studies with mice, it was determined that the insulin inhibitory receptor is responsible for preventing insulin-producing beta cells from participating in regular insulin signalling pathways. The pathways allow insulin to increase glucose uptake in certain tissues and decrease glucose synthesis in the liver to maintain proper blood glucose levels. The activity of the insulin inhibitory receptor is upregulated in T2D patients, which results in insulin resistance (Ansarullah et al., 2021). Since T2D is caused by a decreased ability of tissue cells to respond to insulin, treatments that sensitize muscle, liver and fat cells to insulin by targeting insulin inhibitory receptors can protect the patient from beta cell loss. Using monoclonal antibodies, the activity of the insulin inhibitory receptor was removed in diabetic mice, which increased insulin signalling and decreased T2D symptoms. In the future, research may investigate the targeting of insulin inhibitory receptors in humans as a potential therapeutic approach for T2D patients (Ansarullah et al., 2021).

ARTIFICIAL INTELLIGENCE FOR T1D AND T2D

Artificial intelligence (AI) is a branch of computer science that can create systems for data analysis and management of complex issues. In the context of diabetes, AI can improve disease management for healthcare systems, healthcare professionals and patients. It allows for automated diabetes screening, provides clinical decision support and supplies self-management tools for diabetes patients (Ellahham, 2020).

For healthcare systems, AI provides efficient diabetes management, which decreases the amount of time that a patient spends in a healthcare facility and reduces the need for patients to constantly travel to their healthcare provider. It has improved patient flow and productivity within healthcare facilities. Moreover, telehealth allows for remote monitoring of diabetes, which has replaced 50%-70% of in-person follow-up visits with virtual consultations (Ellahham, 2020).

Diabetes patients are also at risk of diabetic retinopathy, which is an eye condition that leads to blindness. Fortunately, AI is able to provide cost-effective and early detection of diabetic retinopathy by screening images of any changes in the patient's retina. AI is able to provide quite accurate early diagnoses of the eye condition that may otherwise be unnoticed by some healthcare professionals. These algorithms have a retina screening sensitivity of 92.3% and a screening specificity of 93.7%. This allows healthcare professionals to provide timely interventions to decrease retinopathy symptoms (Ellahham, 2020).

Diabetic neuropathy occurs when high blood glucose levels damage the nerves in diabetic patients. The first symptoms of diabetic neuropathy can be identified at the location of small nerve fibres, however, current screening methods for diabetic neuropathy diagnosis do not provide quantification of nerve fibres at the initial site of injury (Williams et al., 2020). Another screening method utilizes corneal confocal microscopy, however, the nerves that are imaged must be distinguished from other cell structures in the background. The process of analyzing images is labour-intensive and requires experts to diagnose the condition, which decreases the productivity in healthcare systems. It is proposed that AI can be used to automate the diagnosis process and quantify nerve fibres in the cornea (Williams et al., 2020). Convolutional neural networks are a branch of machine learning that can classify images of the cornea as either indicators of neuropathy or a healthy individual. It is a non-invasive diagnostic approach that eliminates the need for manual analysis of the corneal images, quantifies the amount of nerve fibers at the site of injury, increases healthcare system efficiency and enables early detection of neuropathy (Williams et al., 2020). In a recent study, 132 individuals affected with neuropathy and 90 individuals without neuropathy were used as test subjects to assess the accuracy of a convolutional neural network algorithm. The diagnostic algorithm had a diagnostic specificity of 80% and a diagnostic sensitivity of 70%. Further developments are ongoing to investigate the use of AI to diagnose diabetic neuropathy (Williams et al., 2020).

Case-based reasoning (CBR) is an AI technique that can solve issues using knowledge learned from similar past experiences. For example, CBR can provide automatic detection of blood glucose control issues, suggest possible solutions and remember the list of successful and unsuccessful solutions for each diabetic patient. This allows for more efficient analysis of a patient's disease progression and produces individualized insulin therapy solutions (Ellahham, 2020).

Machine learning is a branch of AI that can identify people at high risk for diabetes by assessing genetic factors, lifestyle, diet and mental health. Using data from 68,994 non-diabetic and diabetic people, an AI algorithm was able to predict diabetes with an accuracy of over 80%. This is a more effective diagnostic tool compared to monitoring a patient's blood glucose levels (Ellahham, 2020). Additionally, machine learning can also predict the development of both short-term and long-term diabetes-related complications for each patient. This allows healthcare professionals to provide earlier therapies to prevent future complications (Ellahham, 2020).

For T2D, maintaining a proper diet and regularly exercising are two important factors for T2D management. Mobile apps have been designed to provide custom diet and exercise plans that match a patient's lifestyle.

Exercise intensity and duration can be tracked through wearable devices that can motivate the patient to continue their habits (Ellahham, 2020). Additionally, the daily calorie consumption and nutrient details of a patient's meals can be calculated by uploading an image of the food to certain apps. This may help patients monitor their weight and prevent obesity (Ellahham, 2020).

Although AI provides numerous benefits for diabetes management, there are several limitations. AI algorithms require constant refinements in order to keep up with new discoveries in diabetes research and provide accurate solutions. There is also a high cost with the initial development and maintenance of these algorithms, which limits their usage in lower income countries (Ellahham, 2020). Moreover, there may be limitations in the amount of data available for machine learning algorithms that rely on large datasets for accurate diagnoses and solutions. Since the incorporation of AI into healthcare is a relatively new approach to diabetes management, the long-term efficacy and accuracy of these applications has not yet been studied. A continuous use of AI systems may help determine the long-term benefits of algorithms and standardize the use of AI for diabetes management (Ellahham, 2020).

CONCLUSION

Advancements in the diagnosis, management and treatment of T1D and T2D are expected to improve the efficiency of healthcare systems, aid healthcare professionals in decision making and allow patients to self-manage their disease progression (Ellahham, 2020). Alternatives to daily insulin injections and improved transplantation techniques may provide a more pleasant experience in the self-management of diabetes and provide more convenient treatment options (Pathak et al., 2019). Overall, the future of diabetes is focused on providing more comfortable and long-lasting treatment options to improve the quality of life for diabetic patients and improve the productivity of healthcare systems.

REFERENCES

CHAPTER 1:

Ahmed, A. M. (2002). History of diabetes mellitus. Saudi medical journal, 23(4), 373-378.

Barnett, R. (2010). Diabetes. The Lancet, 375(9710), 191.

Luft, R. (1989). Oskar Minkowski: discovery of the pancreatic origin of diabetes, 1889. Diabetologia, 32(7), 399-401.

Karamanou, M., Protogerou, A., Tsoucalas, G., Androutsos, G., & Poulakou-Rebelakou, E. (2016). Milestones in the history of diabetes mellitus: The main contributors. World journal of diabetes, 7(1), 1.

Laios, K., Karamanou, M., Saridaki, Z., & Androutsos, G. (2012). Aretaeus of Cappadocia and the first description of diabetes. Hormones, 11(1), 109-113.

Simoni, R. D., Hill, R. L., & Vaughan, M. (2002). The discovery of insulin: the work of Frederick Banting and Charles Best. Journal of Biological Chemistry, 277(26), e15-e15.

Zimmet, P. Z., Magliano, D. J., Herman, W. H., & Shaw, J. E. (2014). Diabetes: a 21st century challenge. The lancet Diabetes & endocrinology, 2(1), 56-64.

CHAPTER 2:

Basina, M. (2018). *Choosing a glucose meter*. Healthline. https://www.healthline.com/health/type-2-diabetes/choosing-glucose-meter
Christopoulou-Aletra, H., Papavramidou, N. (2008). 'Diabetes' as described by Byzantine writers from the fourth to the ninth century AD: the Graeco-Roman influence. *Diabetologia*. 51: 892-896.

https://doi.org/10.1007/s00125-008-0981-4

Delgado, G. et al. (2018). Dietary Intervention with Oatmeal in Patients with uncontrolled Type 2 Diabetes Mellitus - A Crossover Study. *Exp Clin Endocrinol Diabetes.* 127(9): 623-629. https://doi.org/10.1055/a-0677-6068

Eknoyan, G., Nagy, J. (2005). A history of diabetes mellitus or how a disease of the kidneys evolved into a kidney disease. *Advances in Chronic Kidney Disease.* 12(2): 223–229. https://doi.org/10.1053/j.ackd.2005.01.002

Gordon, A., Buch, Z., Baute, V., Coeytaux, R. (2019). Use of Ayurveda in the Treatment of Type 2 Diabetes Mellitus. Glob Adv Health Med. 8: 2164956119861094. https://doi.org/10.1177/2164956119861094

McCoy, K. (2009). *The history of diabetes.* Everyday Health. https://www.everydayhealth.com/diabetes/understanding/diabetes-mellitus-through-time.aspx

Patlak, M. (2002). New weapons to combat an ancient disease: treating diabetes. *The FASEB Journal.* 16(14): 1853. https://doi.org/10.1096/fj.02-0974bkt

Pollack, A. (2005). *Lizard-Derived Diabetes Drug Is Approved by the F.D.A. T*he New York Times https://www.nytimes.com/2005/04/30/business/lizardderived-diabetes-drug-is-approved-by-the-fda.html

Sanders, J. L. (2002). From Thebes to Toronto and the 21st century: an incredible journey. *Diabetes Spectrum*, 15(1): 56-60. https://doi.org/10.2337/diaspect.15.1.56

Weatherspoon, D. (2020). *Diabetes: past treatments, new discoveries.* Medical News Today. https://www.medicalnewstoday.com/articles/317484

Werner, E., Bell, J. (1922). The preparation of methylguanidine, and of ββ-dimethylguanidine by the interaction of dicyandiamide, and methylammonium and dimethylammonium chlorides respectively. *J. Chem. Soc., Trans.* 121: 1790–95. https://doi.org/10.1039/CT9222101790

CHAPTER 3:

Weatherspoon, D, (2020). Diabetes: Past Treatments, new discoveries. Medical News Today. https://www.medicalnewstoday.com/articles/317484#non-insulin-treatment

White, J. R., (2014). A Biref History of the Development of Diabetes Medications. Diabetes Spectr. 27(2): 82–86. doi: 10.2337/diaspect.27.2.82

Zinman, B., Skyler J. S., Riddle, M. C., Ferrannini, E., (2017). Diabetes Research and Care Through the Ages. Diabetes Care. 40(10): 1302–1313. https://doi.org/10.2337/dci17-0042

Rockefeller University, (2021) An Effective Dietary Therapy for Diabetes Before the Discovery of Insulin. Hospital Centennial. https://centennial.rucares.org/index.php?page=Dietary_Therapy_Diabetes

Mazur, A., (2011). Why were "starvation diets" promoted for diabetes in the pre-insulin period?. Nutr J. 10:23. doi: 10.1186/1475-2891-10-23

Vecchio, I., Tornali, C., Bragazzi, N. L., Martini, M., (2018). The Discovery of Insulin: An Important Milestone in the History of Medicine. Front Endocrinol (Lausanne). 9: 613. doi: 10.3389/fendo.2018.00613

Al-Tabakha, M. M., Arida, A. I., (2008). Recent Challenges in Insulin Delivery Systems: A Review. Indian J Pharm Sci. 70(3): 278–286. doi: 10.4103/0250-474X.42968

Selam, J. L., (2010). Evolution of Diabetes Insulin Delivery Devices. Sage Journals. 4(3): 505–513. https://doi.org/10.1177/193229681000400302

Blum, A., (2018). Freestyle Libre Glucose Monitoring System. Clin Diabetes. 36(2): 203–204. doi: 10.2337/cd17-0130

Al Hayek, A. A., Robert, A. A., Al Darwish, M. A., (2020). Acceptability of the FreeStyle Libre Flash Glucose Monitoring System: The Experience of Young Patients With Type 1 Diabetes. Clin Med Insights Endocrinol Diabetes. 13:1179551420910122. doi: 10.1177/1179551420910122

Boyle, J. O., Honeycutt, A. A., Narayan, V., Hoerger, T. J., Geiss, L. S., Chen, H., Thompson, T. J., (2001). Projection of Diabetes Burden Through 2050. Diabetes Care. 24(11): 1936-1940. https://doi.org/10.2337/diacare.24.11.1936

CHAPTER 4:

Anja, B., & Laura, R. (2017). The cost of diabetes in Canada over 10 years: applying attributable health care costs to a diabetes incidence prediction model. *Health promotion and chronic disease prevention in Canada: research, policy and practice, 37*(2), 49.

Cannon, A., Handelsman, Y., Heile, M., & Shannon, M. (2018). Burden of illness in type 2 diabetes mellitus. *Journal of managed care & specialty pharmacy, 24*(9-a Suppl), S5-S13.

D. (2019, April 8). One in three Canadians is living with diabetes or PRE-DIABETES, yet knowledge of risk and complications of disease remains low. Retrieved May 15, 2021, from https://www.diabetes.ca/media-room/press-releases/one-in-three-canadians-is-living-with-diabetes-or-prediabetes,-yet-knowledge-of-risk-and-complicatio#_ftnref1

Dendup, T., Feng, X., Clingan, S., & Astell-Burt, T. (2018). Environmental risk factors for developing type 2 diabetes mellitus: a systematic review. *International journal of environmental research and public health*, *15*(1), 78.

Einarson, T. R., Acs, A., Ludwig, C., & Panton, U. H. (2018). Prevalence of cardiovascular disease in type 2 diabetes: a systematic literature review of scientific evidence from across the world in 2007–2017. *Cardiovascular diabetology*, *17*(1), 1-19.

Khan, M. A. B., Hashim, M. J., King, J. K., Govender, R. D., Mustafa, H., & Al Kaabi, J. (2020). Epidemiology of type 2 diabetes–global burden of disease and forecasted trends. *Journal of epidemiology and global health*, *10*(1), 107.

Lin, X., Xu, Y., Pan, X., Xu, J., Ding, Y., Sun, X., ... & Shan, P. F. (2020). Global, regional, and national burden and trend of diabetes in 195 countries and territories: an analysis from 1990 to 2025. *Scientific reports*, *10*(1), 1-11.

Mandal, A. (2019, February 27). Obesity and fast food. Retrieved May 15, 2021, from https://www.news-medical.net/health/Obesity-and-Fast-Food.aspx

Marcin, J. (2018, June 13). Most Common Noncommunicable Diseases. *Healthline*. Retrieved from https://www.healthline.com/health/non-communicable-diseases-list

Miranda, T. (n.d.). Daily sugar intake. Retrieved May 15, 2021, from https://www.angelesinstitute.edu/thenightingale/daily-sugar-intake#:~:text=The%20average%20American%20consumes%2017,calories%20per%20day%20for%20women

N. (n.d.). Symptoms & causes of diabetes. Retrieved May 15, 2021, from https://www.niddk.nih.gov/health-information/diabetes/overview/symptoms-causes#causes

Panagiotopoulos, C., Hadjiyannakis, S., & Henderson, M. (2018). Type 2 diabetes in children and adolescents. *Canadian journal of diabetes*, *42*, S247-S254.

Panahi, S., & Tremblay, A. (2018). Sedentariness and health: is sedentary behavior more than just physical inactivity?. *Frontiers in public health*, *6*, 258.

Patterson, R., McNamara, E., Tainio, M., de Sá, T. H., Smith, A. D., Sharp, S. J., ... & Wijndaele, K. (2018). Sedentary behaviour and risk of all-cause, cardiovascular and cancer mortality, and incident type 2 diabetes: a systematic review and dose response meta-analysis. *European journal of epidemiology*, *33*(9), 811-829.

Sami, W., Ansari, T., Butt, N. S., & Ab Hamid, M. R. (2017). Effect of diet on type 2 diabetes mellitus: A review. *International journal of health sciences*, *11*(2), 65.

T. (n.d.). Mortality due to diabetes. Retrieved May 15, 2021, from https://www.conferenceboard.ca/hcp/Details/Health/mortality-diabetes.aspx-#ftn7-ref

Taylor, D. (2013). Physical activity is medicine for older adults. *Postgraduate Medical Journal*, *90*(1059). Retrieved from https://pmj.bmj.com/content/90/1059/26

Team, W. (2020, September 30). Why it's time to start replacing your daily soda. Retrieved May 15, 2021, from https://health.clevelandclinic.org/soda-do-you-drink-it-every-day/

World Health Organization. (2021, April 13). Diabetes. Retrieved May 15,

2021, from World Health Organization website: https://www.who.int/news-room/fact-sheets/detail/diabetes

CHAPTER 5:

Center for Disease Control and Prevention. (2019, June 27). *Diabetes.* Centers for Disease Control and Prevention. https://www.cdc.gov/diabetes/managing/index.html

Diabetes Canada. (2021, 01 01). *What is diabetes?* Diabetes.ca. https://www.diabetes.ca/about-diabetes/what-is-diabetes

Glucagon. (2018, March 1). You and Your Hormones. https://www.yourhormones.info/hormones/glucagon/

Healthline. (2018, October 4). *Everything You Need to Know About Diabetes.* Healthline.com. https://www.healthline.com/health/diabetes#diet

Mayo Clinic. (2020, October 30). *Diabetes.* mayoclinic.org. https://www.mayoclinic.org/diseases-conditions/diabetes/symptoms-causes/syc-20371444

National Institute of Diabetes and Digestive and Kidney Diseases. (2016, December 1). *Diabetes Overview.* niddk.nih.gov. https://www.niddk.nih.gov/health-information/diabetes/overview/symptoms-causes

Wikipedia. (2021, April 18). *Carbohydrate Metabolism.* Wikipedia the Free Encyclopedia. https://en.wikipedia.org/wiki/Carbohydrate_metabolism

CHAPTER 6:

Centers for Disease Control and Prevention. (2019, October 8). *10 Tips for Coping with Diabetes Distress.* Centers for Disease Control and Prevention. https://www.cdc.gov/diabetes/managing/diabetes-distress/

ten-tips-coping-diabetes-distress.html.

EmmaHook. (n.d.). *Diabetes and your emotions*. Diabetes UK. https://www.diabetes.org.uk/guide-to-diabetes/emotions.

Llamas, M. (2016, November 17). *Diabetes - Stigma, Blame and Shame*. Drugwatch.com. https://www.drugwatch.com/featured/diabetes-stigma/.

Solomon, R. (2016, October 12). *I Have Diabetes. Am I to Blame?* The New York Times. https://www.nytimes.com/2016/10/12/opinion/diabetes-diet-and-shame.html.

Tabish S. A. (2007). Is Diabetes Becoming the Biggest Epidemic of the Twenty-first Century?. *International journal of health sciences*, *1*(2), V–VIII.

World Health Organization. (n.d.). *Diabetes*. World Health Organization. https://www.who.int/news-room/fact-sheets/detail/diabetes.

CHAPTER 7:

American Diabetes Association. (2021). Insulin and other Injectables. https://www.diabetes.org/healthy-living/medication-treatments/insulin-other-injectables/insulin-routines

American Diabetes Association. (2019). Diabetes Technology: Standards of Medical Care in Diabetes-2019. *Diabetes Care*. 42(1), 71-80. https://care.diabetesjournals.org/content/diacare/42/Supplement_1/S71.full.pdf

Brown, G. (2020). What is a CGM (Continuous Glucose Monitor) and How do I Choose One. *Healthline*. https://www.healthline.com/diabetesmine/what-is-continuous-glucose-monitor-and-choosing-one

CADTH. (2021). Hybrid Closed-Loop Insulin Delivery Systems for

People with Type 1 Diabetes. *Canadian Journal of Health Technologies.* https://www.cadth.ca/hybrid-closed-loop-insulin-delivery-systems-people-type-1-diabetes

FDA. (2021). FDA Approves First Treatment for Severe Hypoglycemia that can be Administered Without an Injection. (2019). *U.S Food and Drug Administration.* https://www.fda.gov/news-events/press-announcements/fda-approves-first-treatment-severe-hypoglycemia-can-be-administered-without-injection#:~:text=FDA%20News%20Release-,FDA%20approves%20first%20treatment%20for%20severe%20hypoglycemia,be%20administered%20without%20an%20injection&text=The%20U.S.%20Food%20and%20Drug,be%20administered%20without%20an%20injection

Health Canada. (2012). Insulin Pumps. *Government of Canada.* https://www.canada.ca/en/health-canada/services/healthy-living/your-health/medical-information/insulin-pumps.html

Healthwise. (2021). Insulin Syringes. *MyHealth.Alberta.ca.* https://myhealth.alberta.ca/Health/pages/conditions.aspx?hwid=hw39086#:~:text=An%20insulin%20syringe%20has%20three,skin%20easily%20and%20lessen%20pain

Knebel, T and Neumiller, J. (2019). Medtronic MiniMed 670 G Hybrid Closed-Loop System. *Clinical Diabetes.* 37(1), 94-95. https://clinical.diabetesjournals.org/content/37/1/94

Kesavadev, J., Saboo, B., Krishna, M. B., Krishnan, G. (2020). Evolution of Insulin Delivery Devices: From Syringes, Pens, and Pumps to DIY Artificial Pancreas. *Diabetes Therapy.* 11(6), 1251-1269. https://www.ncbi.nlm.nih.gov/pmc/articles/PMC7261311/

Lori, D., Berard, R., Siemens, Woo, V. (2018). Monitoring Glycemic Control. *Diabetes Canada.* https://guidelines.diabetes.ca/cpg/chapter9

Mayo Clinic. (2019). Blood Glucose Meter: How to Choose. https://www.mayoclinic.org/diseases-conditions/diabetes/in-depth/blood-glucose-meter/art-20046335

NIH. (2021). What is Diabetes? *National Institute of Diabetes and Digestive and Kidney Diseases.* https://www.niddk.nih.gov/health-information/diabetes/overview/what-is-diabetes

Pietrzak, I., and Szadkowska, A. (2020). Nasal Glucagon - a New Way to Treat Severe Hypoglycemia in Patients with Diabetes.

CHAPTER 8:

Abdul-Ghani, M. A., & DeFronzo, R. A. (2009). Pathophysiology of prediabetes. *Current diabetes reports, 9,* 193-199. https://link.springer.com/content/pdf/10.1007/s11892-009-0032-7.pdf

Bhansali, A., & Dutta, P. (2005). Pathophysiology of prediabetes. Journal of the Indian Medical Association, 103(11), 594–599. https://pubmed.ncbi.nlm.nih.gov/16570762/#:~:text=The%20pathophysiology%20of%20prediabetes%20is,of%20large%20pulses%20are%20decreased

Bruns, D. E., Metzger, B. E., & Sacks, D. B. (2020). Diagnosis of Gestational Diabetes Mellitus Will Be Flawed until We Can Measure Glucose. *Clinical chemistry, 66(2),* 265–267. https://doi.org/10.1093/clinchem/hvz027

Centers for Disease Control and Prevention. (2020, June 11). *The surprising truth about prediabetes.* Centers for Disease Control and Prevention. https://www.cdc.gov/diabetes/

library/features/truth-about-prediabetes.html#:~:text=Type%201%20oc-curs%20most%20often,diabetes%2C%20but%20not%20type%201

Dansinger, M. (2019, December 13). *Types of diabetes mellitus.* WebMD. https://www.webmd.com/diabetes/guide/types-of-diabetes-mellitus#2-5

Dedrick, S., Sundaresh, B., Huang, Q., Brady, C., Yoo, T., Cronin, C., Rud-nicki, C., Flood, M., Momeni,
B., Ludvigsson, J., & Altindis E. (2020) The role of gut microbiota and en-vironmental factors in type diabetes pathogenesis. *Frontiers in Endocrinol-ogy, 11,* 78. https://www.frontiersin.org/article/10.3389/fendo.2020.00078

Dendup, T., Feng, X., Clingan, S., & Astell-Burt, T. (2018) Environmental risk factors for developing
type 2 diabetes ,ellitus: a systematic review. *International Journal of Envi-ronmental Research and Public Health, 15(1),* 78. https://doi.org/10.3390/ijerph15010078

Esposito, S., Toni, G., Tascini, G., Santi, E., Berioli, M. G., & Principi, N. (2019) Environmental factors
associated with type 1 diabetes. *Frontiers in Endocrinology, 10,* 592. https://www.frontiersin.org/article/10.3389/fendo.2019.00592

Giannakou, K., Evangelou, E., Yiallouros, P., Christophi, C. A., Middleton, N., Papatheodorou, E., &
Papatheodorou, S. I. (2019). Risk factors for gestational diabetes: An um-brella review of meta-analyses of observational studies. *PLOS ONE, 14(4), e0215372.* https://doi.org/10.1371/journal.pone.0215372

Halim, M., & Halim, A. (2019). The effects of inflammation, aging and oxidative stress on the
pathogenesis of diabetes mellitus (type 2 diabetes). *Diabetes & Metabolic Syndrome: Clinical Research & Reviews, 13(2), 1165-1172.* https://doi.

org/10.1016/j.dsx.2019.01.040.

Kolb, H., & Martin, S. (2017) Environmental/lifestyle factors in the patho-genesis and prevention of type 2
diabetes. *BMC Med 15,* 131. https://doi.org/10.1186/s12916-017-0901-x

Mayo Clinic Staff. (2020, September 22). *Prediabetes.* Mayo Clinic.
https://www.mayoclinic.org/diseases-conditions/prediabetes/diagno-sis-treatment/drc-20355284

Najafi, F., Hasani, J., Izadi, N., Hashemi-Nazari, S. S., Namvar, Z., Mo-hammadi, S., & Sadeghi, M.
(2019). The effect of prepregnancy body mass index on the risk of ges-tational diabetes mellitus: A systematic review and dose-response me-ta-analysis. *Obes Rev, 20(3), 472-486.* https://pubmed.ncbi.nlm.nih.gov/30536891/

Paschou, S. A., Papadopoulou-Marketou, N., Chrousos, G. P., & Kana-ka-Gantenbein, C. (2018) On type 1
diabetes mellitus pathogenesis. *Endocr Connect, 7(1):R38-R46.* https://doi.org/10.1530/EC-17-0347

Rooney, M. R, .Rawlings, A. M., Pankow, J. S., Tcheugui, J. B. E., Coresh, J., Sharrett, R., & Selvin, E.
(2021) Risk of progression to diabetes among older adults with pre-diabetes. *JAMA Intern Med, 181(4),* 511–519. doi:10.1001/jamaint-ernmed.2020.8774

The Canadian Diabetes Association. (2021). *Assess your risk of developing diabetes.* Diabetes Canada.
https://www.diabetes.ca/type-2-risks/risk-factors---assessments

The Canadian Diabetes Association. (2018). *Prediabetes.* Diabetes Canada.

https://www.diabetes.ca/DiabetesCanadaWebsite/media/Managing-My-Di-abetes/Tools%20and%20Resources/prediabetes-fact-sheet.pdf?ext=.pdf

WebMD. (2021). *Diabetes health center.* WebMD. https://www.webmd.com/diabetes/default.htm

World Health Organization. (2021, April 13). *Diabetes.* World Health Organization.
https://www.who.int/news-room/fact-sheets/detail/diabetes

Zhao, R. F., Zhang, W. Y., & Zhou, L. (2017) Relationship between the risk of emergency cesarean
section for nullipara with the prepregnancy body mass index or gestational weight gain. *Zhonghua Fu Chan Ke Za Zhi, 52(11), 757-764.* https://pubmed.ncbi.nlm.nih.gov/29179271/

Zhou, Z., Sun, B., Li, X., & Zhu, C. (2018). DNA methylation landscapes in the pathogenesis of type 2
diabetes mellitus. *Nutr Metab (Lond) 15, 47.* https://doi.org/10.1186/s12986-018-0283-x

CHAPTER 9:

Centers for Disease Control and Prevention. (2020). *Risk factors for dia-betes-related complications.* Retrieved May 16, 2021, from   https://www.cdc.gov/diabetes/data/statistics-report/risks-complications.html.
Mayo Clinic. (2020). *Diabetes.* Retrieved May 16, 2021, from https://www.mayoclinic.org/diseases-conditions/diabetes/symptoms-caus-es/syc-20371444

Reid, T. S. (2007). *Insulin for type 2 diabetes mellitus: Separating the myths from the facts.* Insulin, 2(4),   182–189. https://doi-org.libaccess.lib.mcmaster.ca/10.1016/S1557-0843(07)80062-5

Sakurai, M., Nakamura, K., Miura, K., Takamura, T., Yoshita, K., Naga-
sawa, S., Morikawa, Y., Ishizaki, M., Kido, T., Naruse, Y., Suwazono, Y.,
Sasaki, S., & Nakagawa, H. (2014). *Sugar-sweetened beverage and diet
soda consumption and the 7-year risk for type 2 diabetes mellitus in middle-aged Japanese men.* European Journal of Nutrition, 53(1), 251–258.

Stern, D., Mazariegos, M., Ortiz-Panozo, E., Campos, H., Malik, V. S.,
Lajous, M., & López-Ridaura, R. (2019). *Sugar-sweetened soda consumption increases diabetes risk among Mexican women.* Journal of Nutrition,
149(5), 795–803.

CHAPTER 10:

Ansarullah, Jain, C., Far, F. F., Homberg, S., Wißmiller, K., von Hahn, F.
G., Raducanu, A., Schirge, S., Sterr, M., Bilekova, S., Siehler, J., Wiener,
J., Oppenländer, L., Morshedi, A., Bastidas-Ponce, A., Collden, G., Irmler,
M., Beckers, J., Feuchtinger, A.,... Lickert, H. (2021). Inceptor counteracts
insulin signalling in β-cells to control glycaemia. *Nature, 592*(7852), E1.
https://doi.org/10.1038/s41586-021-03347-z

Bajaj, H. S., Bergenstal, R. M., Christoffersen, A., Davies, M. J., Gowda,
A., Isendahl, J., Lingvay, I., Senior, P. A., Silver, R. J., Trevisan, R., &
Rosenstock, J. (2021). Switching to Once-Weekly Insulin Icodec Ver-
sus Once-Daily Insulin Glargine U100 in Type 2 Diabetes Inadequately
Controlled on Daily Basal Insulin: A Phase 2 Randomized Controlled
Trial. *Diabetes care*, dc202877. Advance online publication. https://doi.
org/10.2337/dc20-2877

Ellahham, S. (2020). Artificial Intelligence: The Future for Diabetes
Care. *The American Journal for Medicine, 113*(8), 895–900. https://doi.
org/10.1016/j.amjmed.2020.03.033

Lingvay, I., Buse, J. B., Franek, E., Hansen, M. V., Koefoed, M. M.,
Mathieu, C., Pettus, J., Stachlewska, K., & Rosenstock, J. (2021). A Ran-

domized, Open-Label Comparison of Once-Weekly Insulin Icodec Titration Strategies Versus Once-Daily Insulin Glargine U100. *Diabetes care*, dc202878. Advance online publication. https://doi.org/10.2337/dc20-2878

Pathak, V., Pathak, N. M., O'Neill, C. L., Guduric-Fuchs, J., & Medina, R. J. (2019). Therapies for Type 1 Diabetes: Current Scenario and Future Perspectives. *Clinical medicine insights. Endocrinology and diabetes, 12*, 1179551419844521. https://doi.org/10.1177/1179551419844521

Williams, B. M., Borroni, D., Liu, R., Zhao, Y., Zhang, J., Lim, J., Ma, B., Romano, V., Qi, H., Ferdousi, M., Petropoulos, I. N., Ponirakis, G., Kaye, S., Malik, R. A., Alam, U., & Zheng, Y. (2020). An artificial intelligence-based deep learning algorithm for the diagnosis of diabetic neuropathy using corneal confocal microscopy: a development and validation study. *Diabetologia, 63*(2), 419–430. https://doi.org/10.1007/s00125-019-05023-4